SpringerBriefs in Optimization

SpringerBriefs in Optimization showcases algorithmic and theoretical techniques, case studies, and applications within the broad-based field of optimization. Manuscripts related to the ever-growing applications of optimization in applied mathematics, engineering, medicine, economics, and other applied sciences are encouraged.

More information about this series at http://www.springer.com/series/8918

Xin-She Yang • Xing-Shi He

Mathematical Foundations of Nature-Inspired Algorithms

Springer

Xin-She Yang
School of Science and Technology
Middlesex University
London, UK

Xing-Shi He
College of Science
Xi'an Polytechnic University
Xi'an, China

ISSN 2190-8354 ISSN 2191-575X (electronic)
SpringerBriefs in Optimization
ISBN 978-3-030-16935-0 ISBN 978-3-030-16936-7 (eBook)
https://doi.org/10.1007/978-3-030-16936-7

Mathematics Subject Classification: 80M50, 68Q25, 68W20, 90C26, 90C31, 46N10, 97N40, 60J10, 37N40

This Springer imprint is published by the registered company Springer Nature Switzerland AG.
The registered company address is: Gewerbestrasse 11, 6330 Cham, Switzerland

Preface

Optimization is everywhere, from engineering design and business planning to data mining and industrial applications. However, many optimization problems are very challenging to solve, especially highly nonlinear problems, combinatorial problems and large-scale problems. Traditional optimization struggles to cope such tough problems. In recent years, nature-inspired algorithms have become popular for solving tough problems concerning optimization, data mining and machine learning. Despite their wide applications, their mathematical analysis is weak, and, for most new algorithms, there is no analysis at all. Even if there exist some fragmental theoretical studies, it lacks a unified theoretical framework for analysing such algorithms.

This book will attempt to provide a systematic approach to rigorous mathematical analysis of these algorithms based on solid mathematical foundations, and the analyses will be from different angles and perspectives. There are many good textbooks on the detailed introduction of traditional optimization techniques and new algorithms. In contrast, the introduction of algorithms in this book will be relatively brief and will introduce the most relevant algorithms for the purpose of analysing and understanding nature-inspired algorithms. Thus, the introduction of algorithms includes traditional algorithms such as gradient-based methods and new algorithms such as swarm intelligence-based algorithms. The emphasis will be placed on the latest algorithms and their analyses from a wide spectrum of theories and frameworks. Thus, readers can gain greater insight into the main characteristics of different algorithms and understand how and why they work for solving optimization problems in real-world settings. More specifically, in-depth mathematical analyses will be carried out from different perspectives, including complexity theory, fixed-point theory, dynamical system, self-organization, Bayesian framework, Markov chain framework, filter theory, statistical learning and statistical measures. Such mathematical analyses will provide some insight into relevant algorithms and thus allow readers to see the advantages and disadvantages of different algorithms. As a result, it provides further insight and understanding on the performance of each class of algorithms and thus the proper choice of relevant algorithms for a given class of optimization problems to be solved.

It is worth pointing out that there are still many open problems, despite the efforts by all the researchers in this active research area. Our aims here are twofold: to summarize and to highlight. One main intention is to summarize all the major approaches for analysing algorithms from at least ten different angles and perspectives. The other intention is to highlight the key issues and open problems so as to encourage and inspire more research in this area. We hope that more understanding and insight will be gained by new studies and research in the near future.

Therefore, this brief and timely book can serve as a textbook for many optimization courses at both advanced undergraduate and graduate levels. It can also serve as a timely reference for lecturers, graduates and researchers in optimization, operations research, artificial intelligence, data mining, machine learning, computer science and management sciences.

London, UK Xin-She Yang
Xi'an, China Xing-Shi He
January 2019

Contents

1 Introduction to Optimization 1
1.1 Introduction 1
1.2 Essence of an Algorithm 2
1.3 Unconstrained Optimization 4
1.3.1 Univariate Functions 4
1.3.2 Multivariate Functions 7
1.4 Optimization 10
1.5 Gradient-Based Methods 11
1.5.1 Newton's Method 11
1.5.2 Steepest Descent Method 13
1.5.3 Line Search 15
1.5.4 Conjugate Gradient Method 16
1.5.5 Stochastic Gradient Descent 18
1.5.6 Subgradient Method 19

2 Nature-Inspired Algorithms 21
2.1 A Brief History of Nature-Inspired Algorithms 21
2.2 Genetic Algorithms 23
2.3 Simulated Annealing 25
2.4 Ant Colony Optimization 28
2.5 Differential Evolution 29
2.6 Particle Swarm Optimization 31
2.7 Bees-Inspired Algorithms 33
2.8 Bat Algorithm 34
2.9 Firefly Algorithm 35
2.10 Cuckoo Search 38
2.11 Flower Pollination Algorithm 39
2.12 Other Algorithms 40

3 Mathematical Foundations 41
3.1 Convergence Analysis 41
3.1.1 Rate of Convergence 41

3.1.2 Convergence Analysis of Newton's Method 42
3.2 Stability of an Algorithm 44
3.3 Robustness Analysis 45
3.4 Probability Theory 45
3.4.1 Random Variables 45
3.4.2 Poisson Distribution and Gaussian Distribution 48
3.4.3 Common Probability Distributions 49
3.5 Random Walks and Lévy Flights 52
3.6 Performance Measures 54
3.7 Monte Carlo and Markov Chains 56

4 Mathematical Analysis of Algorithms: Part I 59
4.1 Algorithm Analysis and Insight 59
4.1.1 Characteristics of Nature-Inspired Algorithms 59
4.1.2 What's Wrong with Traditional Algorithms? 61
4.2 Advantages of Heuristics and Metaheuristics 61
4.3 Key Components of Algorithms 62
4.3.1 Deterministic or Stochastic 62
4.3.2 Exploration and Exploitation 62
4.3.3 Role of Components 63
4.4 Complexity 64
4.4.1 Time and Space Complexity 64
4.4.2 Complexity of Algorithms 66
4.5 Fixed Point Theory 66
4.6 Dynamical System 67
4.7 Self-organized Systems 68
4.8 Markov Chain Monte Carlo 69
4.8.1 Biased Monte Carlo 70
4.8.2 Random Walks 71
4.9 No-Free-Lunch Theorems 72

5 Mathematical Analysis of Algorithms: Part II 75
5.1 Swarm Intelligence 75
5.2 Filter Theory 76
5.3 Bayesian Framework and Statistical Analysis 76
5.4 Stochastic Learning 77
5.5 Parameter Tuning and Control 78
5.5.1 Parameter Tuning 79
5.5.2 Parameter Control 80
5.6 Hyper-Optimization 81
5.6.1 A Multiobjective View 81
5.6.2 Self-tuning Framework 82
5.6.3 Self-tuning Firefly Algorithm 83
5.7 Multidisciplinary Perspectives 84
5.8 Future Directions 84

6 Applications of Nature-Inspired Algorithms 87
6.1 Design Optimization in Engineering 87
6.1.1 Design of a Spring 87
6.1.2 Pressure Vessel Design 88
6.1.3 Speed Reducer Design 89
6.1.4 Other Design Problems 90
6.2 Inverse Problems and Parameter Identification 90
6.3 Image Processing 91
6.4 Classification, Clustering and Feature Selection 92
6.5 Travelling Salesman Problem 92
6.6 Vehicle Routing 94
6.7 Scheduling 94
6.8 Software Testing 95
6.9 Deep Belief Networks 95
6.10 Swarm Robots 96

References 99

Index 105

About the Authors

Xin-She Yang is a reader/professor at the School of Science and Technology, Middlesex University London. He is also an elected bye-fellow and supervisor at Cambridge University and a guest chief scientist and distinguished professor at Xi'an Polytechnic University, China. He worked at the National Physical Laboratory as a senior research scientist after obtaining his DPhil in Applied Mathematics at Oxford University. With more 250 publications and more than 30 books, his research has been cited more than 36,000 times (according to Google Scholar) with an h-index of 71. He has been on the prestigious lists of Highly Cited Researchers (2016, 2017 and 2018), according to Web of Science/Clarivate Analytics. He is the chair of IEEE CIS Task Force on Business Intelligence and Knowledge Management. He has given many invited keynote talks at international conferences such as ICCS2015 (Iceland), SIBGRAPI2015 (Brazil), IEEE OIPE2016 (Italy), BDIOT2017 (UK), ICIST2018(UK), CPMMI2019 (Serbia), ICAAI2018 (Spain) and LION2019 (Greece). He has also given tutorials at international conferences such as FedCSIS2011 (Poland), IAMG2011 (Austria), ECTA2015 (Portugal), MOD2017 (Italy) and EANN2018 (UK).

Xing-Shi He is a professor in the College of Science at Xi'an Polytechnic University. He graduated with an MSc in Mathematics from Shaanxi Normal University. He has held many professional positions including the deputy dean of the College of Science at Xi'an Polytechnic University. He was a recipient of Shaanxi Province Distinguished Teaching Achievement Award, Hong Kong SangMa Teaching Award and a few National and Provincial Awards for Scientific Achievements. He has been on the scientific committee of Shaanxi Society for Industrial and Applied Mathematics and Chinese Statistical Association and has been advisor of Shaanxi Mathematical Modelling Committee. He has published more than 130 research papers and authored/edited 8 books. He had more than 15 research projects and has organized special sessions in many international conferences, including the IEEE SSCI2014 (Orlando, USA). His research interests include algorithms, mathematical modelling, nature-inspired computation, statistics and probability and computational intelligence.

Chapter 1
Introduction to Optimization

Optimization is part of many university courses because of its importance in many disciplines and applications such as engineering design, business planning, computer science, data mining, machine learning, artificial intelligence and industries. The techniques and algorithms for optimization are diverse, ranging from the traditional gradient-based algorithms to contemporary swarm intelligence based algorithms. This chapter introduces the fundamentals of optimization and some of the traditional optimization techniques.

1.1 Introduction

Optimization is everywhere, though it can mean different things from different perspectives. From basic calculus, optimization can simply be to find the maximum or minimum of a function $f(x)$ such as $f(x) = x^4 + 2x^2 + 1$ in the real domain $x \in \mathbb{R}$. In this case, we can either use a gradient-based method or simply spot the solution due to the fact that x^4 and x^2 are always non-negative; the minimum values of x^4 and x^2 are zero; thus, $f(x) = x^4+2x^2+1$ has a minimum $f_{\min} = 1$ at $x_* = 0$. This can be confirmed easily by taking the first derivative of $f(x)$ with respect to x, we have

$$f'(x) = 4x^3 + 4x = 4x(x^2 + 1) = 0, \tag{1.1}$$

which has only one real solution $x = 0$ because $x^2 + 1$ cannot be zero for any real number x. The condition $f'(x) = 0$ seems to be sufficient to determine the optimal solution in this case. In fact, this function is convex with only one minimum in the whole real domain.

However, things become more complicated when $f(x)$ is highly nonlinear with multiple optima. For example, if we try to find the maximum value of $f(x) =$

X.-S. Yang, X.-S. He, *Mathematical Foundations of Nature-Inspired Algorithms*, SpringerBriefs in Optimization, https://doi.org/10.1007/978-3-030-16936-7_1

$\text{sinc}(x) = \sin(x)/x$ in the real domain, we can naively use

$$f'(x) = \Big[\frac{\sin(x)}{x}\Big]' = \frac{x\cos(x) - \sin(x)}{x^2} = 0, \tag{1.2}$$

which has an infinite number of solutions for $x \neq 0$. There is no simple formula for these solutions; thus, a numerical method has to be used to calculate these solutions. In addition, even with all the efforts to find these solutions, care has to be taken because the actual global maximum $f_{\max} = 1$ occurs at $x_* = 0$. However, this solution can only be found by taking the limit $x \to 0$, and it is not part of the solutions from the above condition of $f'(x) = 0$. This highlights the potential difficulty for nonlinear problems with multiple optima or multi-modality.

Furthermore, not all functions are smooth. For example, if we try to use $f'(x) = 0$ to find the minimum of

$$f(x) = |x| e^{-\sin(x^2)}, \tag{1.3}$$

we will realize that $f(x)$ is not differentiable at $x = 0$, though the global minimum $f_{\min} = 0$ occurs at $x_* = 0$. In this case, optimization techniques that require the calculation of derivatives will not work.

Problems become more challenging in higher-dimensional spaces. For example, the nonlinear function

$$f(\boldsymbol{x}) = \Big\{\Big[\sum_{i=1}^{n} \sin^2(x_i)\Big] - \exp\Big(-\sum_{i=1}^{n} x_i^2\Big)\Big\} \cdot \exp\Big[-\sum_{i=1}^{n} \sin^2\sqrt{|x_i|}\Big], \tag{1.4}$$

where $-10 \leq x_i \leq 10$ (for $i = 1, 2, \ldots, n$), has the global minimum $f_{\min} = -1$ at $\boldsymbol{x}_* = (0, 0, \ldots, 0)$, but this function is not differentiable at $\boldsymbol{x}_*$.

Therefore, optimization techniques have to be diverse to use gradient information when appropriate, and not to use it when it is not defined or not easily calculated. Though the above nonlinear optimization problems can be challenging to solve, constraints on the search domain and certain independent variables can make the search domain much more complicated, which can consequently make the problem even harder to solve. In addition, sometime, we have to optimize several objective functions instead of just one function, which will in turn make a problem more challenging to solve.

1.2 Essence of an Algorithm

An algorithm is a computational, iterative procedure. For example, Newton's method for finding the roots of a polynomial $p(x) = 0$ can be written as

$$x_{t+1} = x_t - \frac{p(x_t)}{p'(x_t)}, \tag{1.5}$$

where x_t is the approximation at iteration t, and $p'(x)$ is the first derivative of $p(x)$. This procedure typically starts with an initial guess x_0 at $t = 0$.

In most cases, as along as $p' \neq 0$ and x_0 is not too far away from the target solution, this algorithm can work very well. As we do not know the target solution

$$x_* = \lim_{t \to \infty} x_t \tag{1.6}$$

in advance, the initial guess can be an educated guess or a purely random guess. However, if the initial guess is too far way, the algorithm may never reach the final solution or simply fail.

For example, for $p(x) = x^2 + 9x - 10 = (x - 1)(x + 10)$, we know its roots are $x_* = 1$ and $x_* = -10$. We also have $p'(x) = 2x + 9$ and

$$x_{t+1} = x_t - \frac{(x_t^2 + 9x_t - 10)}{2x_t + 9}. \tag{1.7}$$

If we start from $x_0 = 10$, we can easily reach $x_* = 1$ in less than five iterations. If we use $x_0 = 100$, it may take about eight iterations, depending on the accuracy we want. If we start any value $x_0 > 0$, we can only reach $x_* = 1$ and we will never reach the other root $x_* = -10$. If we start with $x_0 = -5$, we can reach $x_* = -10$ in about seven steps with an accuracy of 10^{-9}. However, if we start with $x_0 = -4.5$, the algorithm will simply fail because $p'(x_0) = 2x_0 + 9 = 0$.

This has clearly demonstrated that the final solution will usually depend on where the initial starting point is.

This method can be modified to solve optimization problems. For example, for a single objective function $f(x)$, the minimal and maximal values should occur at stationary points $f'(x) = 0$, which becomes a root-finding problem for $f'(x)$. Thus, the maximum or minimum of $f(x)$ can be found by modifying Newton's method as the following iterative formula:

$$x_{t+1} = x_t - \frac{f'(x_t)}{f''(x_t)}. \tag{1.8}$$

For example, we know that the optimal solution $f(x) = x^2 + 2x + 1$ is $f_{\min} = 0$ at $x_* = -1$. Since $f'(x) = 2x + 2$ and $f''(x) = 2$, we have

$$x_{t+1} = x_t - \frac{2x_t + 2}{2} = -1, \tag{1.9}$$

which means that this method can reach the optimal solution $x_* = -1$ in a single step, starting from any initial point $x_t = a \in \mathbb{R}$. This is a special case because $f(x)$ is a quadratic (a convex function), which belongs to a class of convex optimization [18]. In this case, Newton's method is a very efficient algorithm. In general, $f(x)$ is highly nonlinear and certainly not convex; thus, algorithms may not be efficient and such problems can be challenging to solve.

For a D-dimensional problem with an objective $f(\boldsymbol{x})$ with independent variables $\boldsymbol{x} = (x_1, x_2, \ldots, x_D)$, the above iteration formula can be generalized to a vector form

$$\boldsymbol{x}^{t+1} = \boldsymbol{x}^t - \frac{\nabla f(\boldsymbol{x}^t)}{\nabla^2 f(\boldsymbol{x}^t)}, \tag{1.10}$$

where we have used the notation convention $\boldsymbol{x}^t$ to denote the current solution vector at iteration t (not to be confused with an exponent). Both notations x^t and x_t are commonly used in the literature and in many popular textbooks. We will also use both notations almost interchangeably in this book if no confusion arises.

In general, an algorithm A can be written as

$$\boldsymbol{x}^{t+1} = A(\boldsymbol{x}^t, h_*, p_1, \ldots, p_K), \tag{1.11}$$

which represents the fact that the new solution vector is a function of the existing solution vector $\boldsymbol{x}^t$, some historical best solution h_* during the iteration history and a set of algorithm-dependent parameters $(p_1, p_2, \ldots, p_K)$. The exact function forms will depend on the algorithm, and different algorithms are only different in terms of the function form, number of parameters and the ways of using historical data.

1.3 Unconstrained Optimization

An optimization problem is called unconstrained if it has an objective to be optimized, subject to no constraints. If additional conditions are imposed on the permissible values of independent or decision variables, the problem becomes a constrained optimization problem. Let us first look at the simplest unconstrained function optimization problems.

1.3.1 Univariate Functions

The simplest optimization problem without any constraints is probably the search for the maxima or minima of a univariate function $f(x)$ for $-\infty < x < +\infty$ (or in the whole real domain $\mathbb{R}$), we simply write

$$\text{maximize or minimize} \quad f(x), \quad x \in \mathbb{R}, \tag{1.12}$$

or simply

$$\max \text{ or } \min \quad f(x), \quad x \in \mathbb{R}. \tag{1.13}$$

An optimization problem without any constraints on the decision variable x is often called an unconstrained optimization problem. For unconstrained optimization problems, the optimality often occurs at the critical points given by the stationary condition $f'(x) = 0$. However, this stationary condition is just a necessary condition, but it is not a sufficient condition. If $f'(x_*) = 0$ and $f''(x_*) > 0$, it is a local minimum. Conversely, if $f'(x_*) = 0$ and $f''(x_*) < 0$, then it is a local maximum.

However, if $f'(x_*) = 0$ but $f''(x)$ is indefinite (both positive and negative) when $x \to x_*$, then x_* corresponds to a saddle point. For example, $f(x) = x^3$ has a saddle point $x_* = 0$ because $f'(0) = 0$ but f'' changes sign from $f''(0+) > 0$ to $f''(0-) < 0$.

In general, a function can have multiple stationary points. In order to find the global maximum or minimum, we may have to go through every stationary points, unless the objective function is convex.

It is worth pointing out that the notation argmin or argmax is used in some textbooks; thus, the above optimization can be written as

$$\text{argmax}_{x \in \mathbb{R}} f(x), \tag{1.14}$$

or

$$\text{argmin}_{x \in \mathbb{R}} f(x). \tag{1.15}$$

This notation puts its emphasis on the argument x so that the optimization task is to find a point (or points) in the domain of $f(x)$ that maximizes (or minimizes) the function values. On the other hand, the notation we used in (1.12) emphasizes the maximum or minimum value of the objective function. Both types of notations are used in the literature.

For $f(x) = 3x^4 - 20x^3 - 24x^2 + 240x + 400$, where $-\infty < x < +\infty$, its stationary or optimality condition is

$$f'(x) = 12x^3 - 60x^2 - 48x + 240 = 0,$$

which seems not easy to solve analytically. However, we can rewrite it as

$$f'(x) = 12(x + 2)(x - 2)(x - 5) = 0.$$

Thus, there are three solutions $x_* = -2, +2$ and $+5$. The second derivative is

$$f''(x) = 36x^2 - 120x - 48.$$

At $x = 2$, we have $f(2) = 672$ and $f''(2) = -144$; thus, this point is a local maximum.

At $x = 5$, we have $f(5) = 375$ and $f''(5) = 252$, which means that this point is a local minimum.

On the other hand, at $x = -2$, we have $f''(-2) = 336$ and $f(-2) = 32$; thus, this point is a local minimum. Comparing the two minima at $x_* = -2$ and $x_* = 5$, we can conclude that the global minimum occurs at $x_* = -2$ with $f_{\min} = 32$.

As there is one local maximum at $x_* = 2$, can we conclude that the global maximum is $f_{\max} = 672$? The answer is no. If we look any other points such as $x = 7$ or $x = -5$, we have $f(7) = 1247$ and $f(-5) = 2975$, which are much larger than 672; thus, 672 cannot be a global maximum. In fact, this function is unbounded and thus its maximum is $+\infty$. This kind of function is often referred to be as multimodal.

However, if we impose a simple interval such that $x \in [-3, 7]$. Then, we can conclude that the global maximum occurs at $x = 7$ (the right boundary) with $f(7) = 1247$. However, in this case, the maximum does not occur at a stationary point.

This example clearly shows that care should be taken when dealing with multimodal functions.

The maximization of a function $f(x)$ can be converted into the minimization of $A - f(x)$, where A is usually a positive number (though $A = 0$ will do). For example, we know that the maximum of $f(x) = e^{-x^2}$, for $x \in (-\infty, \infty)$, is 1 at $x_* = 0$. This problem can be converted to a minimization problem $-f(x)$. For this reason, the optimization problems can be expressed as either minimization or maximization, depending on the context and convenience of formulations.

For a simple function optimization problem with one independent variable, the mathematical principle may be easy to understand, and the optimality occurs either at $f'(x) = 0$ (stationary points) or at boundaries (limits of simple intervals). However, it may not be so easy to find the actual optimal solution, even for seemingly simple functions such as $\text{sinc}(x) = \sin(x)/x$.

For $f(x) = \frac{\sin(x)}{x}$, it has an infinite number of local minima and maximum with the global maximum $f_{\max} = 1$ occurring at $x = 0$ as seen in Figure 1.1. However, the analytical method may not be so straightforward.

We know that

$$f'(x) = \frac{\cos(x)}{x} - \frac{\sin(x)}{x^2} = 0, \quad (x \neq 0),$$

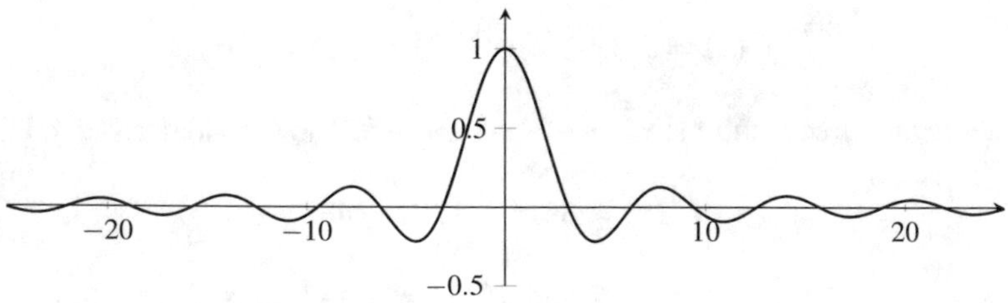

Fig. 1.1 An example of highly nonlinear multimodal functions

which leads to

$$\tan(x) = x, \quad (x \neq 0).$$

But there is no explicit formula for the solutions, except for some approximations. Even we can solve it numerically to find some roots, but we cannot find all the roots (infinitely many).

In addition, these roots do not give any clear indication that $x = 0$ corresponds to the global maximum. In fact, the maximum at $x = 0$ can only be obtained by other methods such as taking limit of an alternative form or by some complex integrals.

As we have seen from this example, numerical methods have to be used to find the actual optimal points. This highlights a key issue: Even we know the basic theory of optimization, it may not directly help much in solving certain classes of problems such as highly nonlinear, multimodal optimization problems. In fact, analytical methods can solve only a small set of problems. For a vast majority of problems, numerical algorithms become essential.

1.3.2 Multivariate Functions

For a multivariate function $f(\boldsymbol{x})$ with n variables where $\boldsymbol{x} = (x_1, \ldots, x_n)^T$, its optimization can be expressed in a similar way to a univariate optimization problem.

$$\text{minimize/maximize} \quad f(\boldsymbol{x}), \quad \boldsymbol{x} \in \mathbb{R}^n. \tag{1.16}$$

Here we have used the notation $\mathbb{R}^n$ to denote that the vector $\boldsymbol{x}$ is in an n-dimensional space where each component x_i is a real number. That is, $-\infty < x_i < +\infty$ for $i = 1, 2, \ldots, n$.

For a function $f(\boldsymbol{x})$, we can expand it locally (with a perturbation $|\epsilon| \ll 1$) using Taylor series about a point $\boldsymbol{x} = \boldsymbol{x}_*$ so that $\boldsymbol{x} = \boldsymbol{x}_* + \epsilon \boldsymbol{u}$

$$f(\boldsymbol{x} + \epsilon \boldsymbol{u}) = f(\boldsymbol{x}_*) + \epsilon \boldsymbol{G}^T(\boldsymbol{x}_*)\boldsymbol{u} + \frac{1}{2}\epsilon^2 \boldsymbol{u}^T \boldsymbol{H}(\boldsymbol{x}_*)\boldsymbol{u} + \ldots, \tag{1.17}$$

where $\boldsymbol{G}$ and $\boldsymbol{H}$ are its gradient vector and Hessian matrix, respectively. ϵ is a small parameter, and $\boldsymbol{u}$ is a local perturbation vector. Here, the superscript T means the transpose of a vector, which converts a column vector into a corresponding row vector and vice versa. For example, for a generic quadratic function

$$f(\boldsymbol{x}) = \frac{1}{2}\boldsymbol{x}^T \boldsymbol{A}\boldsymbol{x} + \boldsymbol{k}^T \boldsymbol{x} + b,$$

where $\boldsymbol{A}$ is a constant square matrix, $\boldsymbol{k}$ is the gradient vector and b is a constant, we have

$$f(\boldsymbol{x}_* + \epsilon \boldsymbol{u}) = f(\boldsymbol{x}_*) + \epsilon \boldsymbol{k}^T \boldsymbol{u} + \frac{1}{2}\epsilon^2 \boldsymbol{u}^T \boldsymbol{A}\boldsymbol{u} + \ldots, \tag{1.18}$$

where

$$f(\boldsymbol{x}_*) = \frac{1}{2}\boldsymbol{x}_*^T \boldsymbol{A}\boldsymbol{x}_* + \boldsymbol{k}^T \boldsymbol{x}_* + b. \tag{1.19}$$

Thus, in order to study the local behaviour of a quadratic function, we mainly need to study $\boldsymbol{G}$ and $\boldsymbol{H}$. In addition, for simplicity, we can take $b = 0$ as it is a constant anyway.

At a stationary point $\boldsymbol{x}_*$, the first derivatives are zero or $\boldsymbol{G}(\boldsymbol{x}_*) = 0$; therefore, Equation (1.17) becomes

$$f(\boldsymbol{x}_* + \epsilon \boldsymbol{u}) \approx f(\boldsymbol{x}_*) + \frac{1}{2}\epsilon^2 \boldsymbol{u}^T \boldsymbol{H}\boldsymbol{u}. \tag{1.20}$$

If $\boldsymbol{H} = \boldsymbol{A}$, then

$$\boldsymbol{A}\boldsymbol{v} = \lambda \boldsymbol{v} \tag{1.21}$$

forms an eigenvalue problem. For an $n \times n$ matrix $\boldsymbol{A}$, there will be in general n eigenvalues $\lambda_j (j = 1, \ldots, n)$ with n corresponding eigenvectors $\boldsymbol{v}_j$.

If $\boldsymbol{A}$ is symmetric, these eigenvectors are either orthonormal or can be converted to be orthonormal. That is,

$$\boldsymbol{v}_i^T \boldsymbol{v}_j = \delta_{ij}, \tag{1.22}$$

where $\delta_{ij} = 1$ if $i = j$, or $\delta_{ij} = 0$ if $i \neq j$.

Near any stationary point $\boldsymbol{x}_*$, if we take $\boldsymbol{u}_j = \boldsymbol{v}_j$ as the local coordinate systems, we then have approximately

$$f(\boldsymbol{x}_* + \epsilon \boldsymbol{v}_j) = f(\boldsymbol{x}_*) + \frac{1}{2}\epsilon^2 \lambda_j, \tag{1.23}$$

which means that the variations of $f(\boldsymbol{x})$, when $\boldsymbol{x}$ moves away from the stationary point $\boldsymbol{x}_*$ along the direction $\boldsymbol{v}_j$, are characterized by the eigenvalues. If $\lambda_j > 0$, $|\epsilon| > 0$ will lead to $|\Delta f| = |f(\boldsymbol{x}) - f(\boldsymbol{x}_*)| > 0$. In other words, $f(\boldsymbol{x})$ will increase as $|\epsilon|$ increases. Conversely, if $\lambda_j < 0$, $f(\boldsymbol{x})$ will decrease as $|\epsilon| > 0$ increases. Obviously, in the special case $\lambda_j = 0$, the function $f(\boldsymbol{x})$ will remain constant along the corresponding direction of $\boldsymbol{v}_j$.

Let us look at an example. We know that the function

$$f(x, y) = xy$$

has a saddle point at (0, 0). It increases along the $x = y$ direction and decreases along $x = -y$ direction. From the above analysis, we know that $\boldsymbol{x}_* = (x_*, y_*)^T = (0, 0)^T$ and $f(x_*, y_*) = 0$. We now have

$$f(\boldsymbol{x}_* + \epsilon \boldsymbol{u}) \approx f(\boldsymbol{x}_*) + \frac{1}{2}\epsilon^2 \boldsymbol{u}^T \boldsymbol{A} \boldsymbol{u},$$

where

$$\boldsymbol{A} = \nabla^2 f(\boldsymbol{x}_*) = \begin{pmatrix} \frac{\partial^2 f}{\partial x^2} & \frac{\partial^2 f}{\partial x \partial y} \\ \frac{\partial^2 f}{\partial x \partial y} & \frac{\partial^2 f}{\partial y^2} \end{pmatrix} = \begin{pmatrix} 0 & 1 \\ 1 & 0 \end{pmatrix}.$$

The eigenvalue problem is simply

$$\boldsymbol{A} \boldsymbol{v}_j = \lambda_j \boldsymbol{v}_j, \quad (j = 1, 2),$$

or

$$\begin{vmatrix} 0 - \lambda_j & 1 \\ 1 & 0 - \lambda_j \end{vmatrix} = 0,$$

whose solutions are

$$\lambda_j = \pm 1.$$

For $\lambda_1 = 1$, the corresponding eigenvector is

$$\boldsymbol{v}_1 = \begin{pmatrix} 1 \\ 1 \end{pmatrix}.$$

Similarly, for $\lambda_2 = -1$, the eigenvector is

$$\boldsymbol{v}_2 = \begin{pmatrix} 1 \\ -1 \end{pmatrix}.$$

Since $\boldsymbol{A}$ is symmetric, $\boldsymbol{v}_1$ and $\boldsymbol{v}_2$ are orthonormal because

$$\boldsymbol{v}_1^T \boldsymbol{v}_2 = 1 \times 1 + 1 \times (-1) = 0.$$

Thus, we have

$$f(\boldsymbol{x}_* + \epsilon \boldsymbol{v}_j) = \frac{1}{2}\epsilon^2 \lambda_j, \quad (j = 1, 2). \tag{1.24}$$

As $\lambda_1 = 1$ is positive, f increases along the direction $\boldsymbol{v}_1 = (1 \quad 1)^T$ which is indeed along the line $x = y$.

Similarly, for $\lambda_2 = -1$, f will decrease along $\boldsymbol{v}_2 = (1 \quad -1)^T$ which is exactly along the line $x = -y$. As there is no zero eigenvalue, the function will not remain constant in the region around (0, 0).

This clearly shows that the properties of local Hessian matrices can indicate how the function may vary in that neighborhood, and thus should provide enough information about its local optimality.

1.4 Optimization

In general, an optimization problem with n decision variables can be formulated as the following constrained optimization problem:

$$\text{minimize} \quad f(\boldsymbol{x}), \quad \boldsymbol{x} = (x_1, x_2, \ldots, x_n)^T \in \mathbb{R}^n, \tag{1.25}$$

subject to

$$h_i(\boldsymbol{x}) = 0, \quad (i = 1, 2, \ldots, I), \quad g_j(\boldsymbol{x}) \leq 0, \quad (j = 1, 2, \ldots, J), \tag{1.26}$$

where h_i and g_j are the equality constraints and inequality constraints, respectively. In most cases, the problem functions $f(\boldsymbol{x})$, $h_i(\boldsymbol{x})$ and $g_j(\boldsymbol{x})$ are all nonlinear, and such nonlinear optimization problems can be challenging to solve. In case of all problem functions are linear, it becomes a linear programming problem. In addition, if h_i and g_j are linear, forming a convex domain, and the objective f is convex, this becomes a convex optimization problem. Both linear programs and convex optimization can have efficient algorithms to solve [18, 28, 121].

Even for linear programs, if all of the variables only take integer values, such problems become integer programming problems, which can be much harder to solve. If some of the variables are discrete or integers, while other values are continuous, it becomes a mixed integer programming (MIP) problem that can also be challenging to solve. In the extreme cases such as the travelling sales problem, there are no efficient algorithms to deal with such problems at all.

There are a wide class of optimization techniques, including linear programming, quadratic programming, convex optimization, interior-point method, trust-region method, conjugate gradient and many others [18, 98, 113, 121]. In general, optimization problems are highly nonlinear and multimodal; these traditional algorithms often struggle to deal with nonlinear problems, any solutions can be found tend to be local optimal solutions. There are no effective methods to find global optimal solutions for highly nonlinear problems with complex constraints. In such cases, sophisticated nature-inspired optimization algorithms tend to be promising alternatives in practice.

Before we introduce nature-inspired algorithms later, let us first focus on some traditional algorithms.

1.5 Gradient-Based Methods

Gradient-based methods are iterative techniques that extensively use the information of the gradient of the objective function during iterations. For the minimization of a function $f(\boldsymbol{x})$, the essence of this method is

$$\boldsymbol{x}^{(k+1)} = \boldsymbol{x}^{(k)} + \alpha g(\nabla f, \boldsymbol{x}^{(k)}), \tag{1.27}$$

where α is the step size which can vary during iterations, and k is the iteration counter. In addition, $g(\nabla f, \boldsymbol{x}^{(k)})$ is a function of the gradient ∇f and the current location $\boldsymbol{x}^{(k)}$, and the search direction is usually along the negative gradient direction (or $-\nabla f$) for minimization problems. Different methods use different forms of $g(\nabla f, \boldsymbol{x}^{(k)})$.

1.5.1 Newton's Method

Newton's method is a root-finding algorithm, but it can be modified for solving optimization problems. This is because optimization is equivalent to finding the root of the first derivative $f'(\boldsymbol{x})$ based on the stationary conditions once the objective function $f(\boldsymbol{x})$ is given. For a continuously differentiable function $f(\boldsymbol{x})$, we have the Taylor expansion in terms of $\Delta \boldsymbol{x} = \boldsymbol{x} - \boldsymbol{x}_k$ about a fixed point $\boldsymbol{x}_k$

$$f(\boldsymbol{x}) = f(\boldsymbol{x}_k) + (\nabla f(\boldsymbol{x}_k))^T \Delta \boldsymbol{x} + \frac{1}{2}(\Delta \boldsymbol{x})^T \nabla^2 f(\boldsymbol{x}_k) \Delta \boldsymbol{x} + \ldots,$$

whose third term is a quadratic form. Hence $f(\boldsymbol{x})$ is minimized if $\Delta \boldsymbol{x}$ is the solution of the following linear equation (after taking the first derivative with respect to the increment vector $\Delta \boldsymbol{x} = \boldsymbol{x} - \boldsymbol{x}_k$):

$$\nabla f(\boldsymbol{x}_k) + \nabla^2 f(\boldsymbol{x}_k) \Delta \boldsymbol{x} = 0, \tag{1.28}$$

which gives

$$\Delta \boldsymbol{x} = -\frac{\nabla f(\boldsymbol{x}_k)}{\nabla^2 f(\boldsymbol{x}_k)}. \tag{1.29}$$

This can be rewritten as

$$\boldsymbol{x} = \boldsymbol{x}_k - \boldsymbol{H}^{-1} \nabla f(\boldsymbol{x}_k), \tag{1.30}$$

where $\boldsymbol{H}^{-1}(\boldsymbol{x}_k)$ is the inverse of the Hessian matrix $\boldsymbol{H} = \nabla^2 f(\boldsymbol{x}_k)$, which is defined as

$$\boldsymbol{H}(\boldsymbol{x}) \equiv \nabla^2 f(\boldsymbol{x}) \equiv \begin{pmatrix} \frac{\partial^2 f}{\partial x_1^2} & \cdots & \frac{\partial^2 f}{\partial x_1 \partial x_n} \\ \vdots & \ddots & \vdots \\ \frac{\partial^2 f}{\partial x_n \partial x_1} & \cdots & \frac{\partial^2 f}{\partial {x_n}^2} \end{pmatrix}. \tag{1.31}$$

This matrix is symmetric due to the fact that

$$\frac{\partial^2 f}{\partial x_i \partial x_j} = \frac{\partial^2 f}{\partial x_j \partial x_i}. \tag{1.32}$$

As each step is an approximation, the next solution $\boldsymbol{x}_{k+1}$ is approximately

$$\boldsymbol{x}_{k+1} = \boldsymbol{x}_k - \boldsymbol{H}^{-1} \nabla f(\boldsymbol{x}_k), \tag{1.33}$$

or in different notations as

$$\boldsymbol{x}^{(k+1)} = \boldsymbol{x}^{(k)} - \boldsymbol{H}^{-1} \nabla f(\boldsymbol{x}^{(k)}). \tag{1.34}$$

If the iteration procedure starts from the initial vector $\boldsymbol{x}^{(0)}$, usually a guessed point in the feasible region, then Newton's formula for the kth iteration becomes

$$\boldsymbol{x}^{(k+1)} = \boldsymbol{x}^{(k)} - \boldsymbol{H}^{-1}(\boldsymbol{x}^{(k)}) \nabla f(\boldsymbol{x}^{(k)}). \tag{1.35}$$

It is worth pointing out that if $f(\boldsymbol{x})$ is quadratic, then the solution can be found exactly in a single step.

In order to speed up the convergence, we can use a smaller step size $\alpha \in (0, 1]$ and we have the modified Newton's method

$$\boldsymbol{x}^{(k+1)} = \boldsymbol{x}^{(k)} - \alpha \boldsymbol{H}^{-1}(\boldsymbol{x}^{(k)}) \nabla f(\boldsymbol{x}^{(k)}). \tag{1.36}$$

Sometimes, it might be time-consuming to calculate the Hessian matrix for second derivatives, especially when the dimensionality is high. A good alternative is to use an $n \times n$ identity matrix $\boldsymbol{I}$ to approximate $\boldsymbol{H}$ so that $\boldsymbol{H}^{-1} = \boldsymbol{I}$, and we have a simple form of the class of quasi-Newton methods

$$\boldsymbol{x}^{(k+1)} = \boldsymbol{x}^{(k)} - \alpha \boldsymbol{I} \nabla f(\boldsymbol{x}^{(k)}) = \boldsymbol{x}^{(k)} - \alpha \nabla f(\boldsymbol{x}^{(k)}), \tag{1.37}$$

which is usually called the steepest descent method for minimization problems. Here, the step size α is also called the learning rate in the literature.

1.5.2 Steepest Descent Method

The essence of this method is to find the lowest possible value of the objective function $f(\boldsymbol{x})$ from the current point $\boldsymbol{x}^{(k)}$. From the Taylor expansion of $f(\boldsymbol{x})$ about $\boldsymbol{x}^{(k)}$, we have

$$f(\boldsymbol{x}^{(k+1)}) = f(\boldsymbol{x}^{(k)} + \Delta\boldsymbol{s}) \approx f(\boldsymbol{x}^{(k)}) + (\nabla f(\boldsymbol{x}^{(k)}))^T \Delta\boldsymbol{s}, \tag{1.38}$$

where $\Delta\boldsymbol{s} = \boldsymbol{x}^{(k+1)} - \boldsymbol{x}^{(k)}$ is the increment vector. Since we try to find a lower (better) approximation to the objective function, it requires that the second term on the right hand is negative. That is,

$$f(\boldsymbol{x}^{(k)} + \Delta\boldsymbol{s}) - f(\boldsymbol{x}^{(k)}) = (\nabla f)^T \Delta\boldsymbol{s} < 0. \tag{1.39}$$

From vector analysis, we know that the inner product $\boldsymbol{u}^T\boldsymbol{v}$ of two vectors $\boldsymbol{u}$ and $\boldsymbol{v}$ is largest when they are parallel (but in opposite directions if larger negative is sought). Therefore, $(\nabla f)^T \Delta\boldsymbol{s}$ becomes the largest descent when

$$\Delta\boldsymbol{s} = -\alpha \nabla f(\boldsymbol{x}^{(k)}), \tag{1.40}$$

where $\alpha > 0$ is the step size. This is the case when the direction $\Delta\boldsymbol{s}$ is along the steepest descent in the negative gradient direction. As we have seen earlier, this method is a quasi-Newton method.

The choice of the step size α is very important. A very small step size means slow movement towards the local minimum, while a large step may overshoot and subsequently makes it move far away from the local minimum. Therefore, the step size $\alpha = \alpha^{(k)}$ should be different at each iteration step and should be chosen so that it minimizes the objective function $f(\boldsymbol{x}^{(k+1)}) = f(\boldsymbol{x}^{(k)}, \alpha^{(k)})$. In each iteration, the gradient and step size will be calculated. Again, a good initial guess of both the starting point and the step size is useful.

Let us minimize the function

$$f(x_1, x_2) = 10x_1^2 + 5x_1x_2 + 10(x_2 - 3)^2,$$

where

$$(x_1, x_2) = [-10, 10] \times [-15, 15],$$

using the steepest descent method starting with the initial $\boldsymbol{x}^{(0)} = (10, 15)^T$. We know the gradient

$$\nabla f = (20x_1 + 5x_2, \;\; 5x_1 + 20x_2 - 60)^T,$$

therefore

$$\nabla f(\boldsymbol{x}^{(0)}) = (275, \quad 290)^T.$$

In the first iteration, we have

$$\boldsymbol{x}^{(1)} = \boldsymbol{x}^{(0)} - \alpha_0 \begin{pmatrix} 275 \\ 290 \end{pmatrix}.$$

The step size α_0 should be chosen such that $f(\boldsymbol{x}^{(1)})$ is at the minimum, which means that

$$f(\alpha_0) = 10(10 - 275\alpha_0)^2$$

$$+5(10 - 275\alpha_0)(15 - 290\alpha_0) + 10(12 - 290\alpha_0)^2$$

should be minimized. This becomes an optimization problem for a single independent variable α_0. All the techniques for univariate optimization problems such as Newton's method can be used to find α_0. We can also obtain the solution by setting

$$\frac{df}{d\alpha_0} = -159725 + 3992000\alpha_0 = 0,$$

whose solution is $\alpha_0 \approx 0.04001$.

At the second step, we have

$$\nabla f(\boldsymbol{x}^{(1)}) = (-3.078, 2.919)^T, \quad \boldsymbol{x}^{(2)} = \boldsymbol{x}^{(1)} - \alpha_1 \begin{pmatrix} -3.078 \\ 2.919 \end{pmatrix}.$$

The minimization of $f(\alpha_1)$ gives $\alpha_1 \approx 0.066$, and the new location of the steepest descent is

$$\boldsymbol{x}^{(2)} \approx (-0.797, 3.202)^T.$$

At the third iteration, we have

$$\nabla f(\boldsymbol{x}^{(2)}) = (0.060, 0.064)^T, \quad \boldsymbol{x}^{(3)} = \boldsymbol{x}^{(2)} - \alpha_2 \begin{pmatrix} 0.060 \\ 0.064 \end{pmatrix}.$$

The minimization of $f(\alpha_2)$ leads to $\alpha_2 \approx 0.040$, and we have

$$\boldsymbol{x}^{(3)} \approx (-0.8000299, 3.20029)^T.$$

Then, the iterations continue until a prescribed tolerance is met.

From calculus, we know that we can set the first partial derivatives equal to zero

$$\frac{\partial f}{\partial x_1} = 20x_1 + 5x_2 = 0, \quad \frac{\partial f}{\partial x_2} = 5x_1 + 20x_2 - 60 = 0,$$

and we know that the minimum occurs exactly at

$$\boldsymbol{x}_* = (-4/5, 16/5)^T = (-0.8, 3.2)^T.$$

The steepest descent method gives almost the exact solution after only three iterations.

In finding the step size α_k in the above steepest descent method, we have used the stationary condition $df(\alpha_k)/d\alpha_k = 0$. Well, you may say that if we use this stationary condition for $f(\alpha_0)$, why not use the same method to get the minimum point of $f(\boldsymbol{x})$ in the first place. There are two reasons here. The first reason is that this is a simple example for demonstrating how the steepest descent method works. The second reason is that even for complicated multiple variables $f(x_1, \ldots, x_n)$ (say $n = 500$), then $f(\alpha_k)$ at any step k is still a univariate function, and the optimization of such $f(\alpha_k)$ is much simpler compared with the original multivariate problem.

It is worth pointing out that in many cases, we do not need to explicitly calculate the step size α_k, and we can instead use a sufficient small value or an adaptive scheme, which can work sufficiently well in practice. In fact, α_k can be considered as a hyper-parameter, which can be either tuned or set adaptively. This point will become clearer when we discuss the line search and stochastic gradient method later in this chapter.

From our example, we know that the convergence from the second iteration to the third iteration is slow. In fact, the steepest descent is typically slow once the local minimization is near. This is because near the local minima, the gradient is nearly zero, and thus, the rate of descent is also slow. If high accuracy is needed near the local minimum, other local search methods should be used.

There are many variations of the steepest descent methods. If the optimization is to find the maximum, then this method becomes the *hill-climbing* method because the aim is to climb up the hill (along the gradient direction) to the highest peak.

1.5.3 Line Search

In the steepest descent method, there are two important parts: the descent direction and the step size (or how far to descend). The calculations of the exact step size may be very time-consuming. In reality, we intend to find the right descent direction. Then a reasonable amount of descent, not necessarily the exact amount, during each iteration will usually be sufficient. For this, we essentially use a line search method.

To find the local minimum of the objective function $f(\boldsymbol{x})$, we try to search along a descent direction $\boldsymbol{s}_k$ with an adjustable step size α_k so that

$$f(\boldsymbol{x}_k + \alpha_k \boldsymbol{s}_k) \tag{1.41}$$

decreases as much as possible, depending on the value of α_k. Loosely speaking, a reasonably right step size should satisfy the Wolfe's conditions [18, 98]:

$$f(\boldsymbol{x}_k + \alpha_k \boldsymbol{s}_k) \leq f(\boldsymbol{x}_k) + \gamma_1 \alpha_k \boldsymbol{s}_k^T \nabla f(\boldsymbol{x}_k), \tag{1.42}$$

and

$$\boldsymbol{s}_k^T \nabla f(\boldsymbol{x}_k + \alpha_k \boldsymbol{s}_k) \geq \gamma_2 \boldsymbol{s}_k^T \nabla f(\boldsymbol{x}_k), \tag{1.43}$$

where $0 < \gamma_1 < \gamma_2 < 1$ are algorithm-dependent parameters. The first condition is a sufficient decrease condition for α_k, often called the Armijo condition or rule, while the second inequality is often referred to as the curvature condition. For most functions, we can use $\gamma_1 = 10^{-4}$–10^{-2}, and $\gamma_2 = 0.1$–0.9. These conditions are usually sufficient to ensure that the algorithm converges in most cases; however, more strong conditions may be needed for some tough functions. The basic steps of the line search method can be summarized in Algorithm 1.

1.5.4 Conjugate Gradient Method

The method of conjugate gradient belongs to a wider class of the so-called Krylov subspace iteration methods. The conjugate gradient method was pioneered by Magnus Hestenes, Eduard Stiefel and Cornelius Lanczos in the 1950s [18, 86]. It was named as one of the top ten algorithms of the twentieth century.

The conjugate gradient method can be used to solve a linear system

$$\boldsymbol{A}\boldsymbol{u} = \boldsymbol{b}, \tag{1.44}$$

Initial guess $\boldsymbol{x}_0$ at $k = 0$;
while $(||\nabla f(\boldsymbol{x}_k)|| > \textit{accuracy})$ **do**
    Find the search direction $\boldsymbol{s}_k = -\nabla f(\boldsymbol{x}_k)$;
    Solve for α_k by decreasing $f(\boldsymbol{x}_k + \alpha \boldsymbol{s}_k)$ significantly;
        while satisfying the Wolfe's conditions;
    Update the solution by $\boldsymbol{x}_{k+1} = \boldsymbol{x}_k + \alpha_k \boldsymbol{s}_k$;
    $k = k + 1$;
end

Algorithm 1: The Basic Steps of the Line Search Method

where $\boldsymbol{A}$ is often a symmetric positive definite matrix. The above system is equivalent to minimizing the following function $f(\boldsymbol{u})$:

$$f(\boldsymbol{u}) = \frac{1}{2}\boldsymbol{u}^T \boldsymbol{A}\boldsymbol{u} - \boldsymbol{b}^T \boldsymbol{u} + v, \tag{1.45}$$

where v is a constant and can be taken to be zero. By differentiating the above function with respect to $\boldsymbol{u}$, we can see that $\nabla f(\boldsymbol{u}) = 0$ leads to $\boldsymbol{A}\boldsymbol{u} = \boldsymbol{b}$.

In general, the size of $\boldsymbol{A}$ can be very large and sparse with $n > 100{,}000$, but it is not required that $\boldsymbol{A}$ is strictly symmetric positive definite. In fact, the main condition is that $\boldsymbol{A}$ should be a normal matrix. A square matrix $\boldsymbol{A}$ is called normal if $\boldsymbol{A}^T\boldsymbol{A} = \boldsymbol{A}\boldsymbol{A}^T$. Therefore, a symmetric matrix is a normal matrix, so is an orthogonal matrix because an orthogonal matrix $\boldsymbol{Q}$ satisfying $\boldsymbol{Q}\boldsymbol{Q}^T = \boldsymbol{Q}^T\boldsymbol{Q} = \boldsymbol{I}$.

The theory behind these iterative methods is closely related to the Krylov subspace $\mathscr{K}_k$ spanned by $\boldsymbol{A}$ and $\boldsymbol{b}$ as defined by

$$\mathscr{K}_k(\boldsymbol{A}, \boldsymbol{b}) = \{\boldsymbol{I}\boldsymbol{b}, \boldsymbol{A}\boldsymbol{b}, \boldsymbol{A}^2\boldsymbol{b}, \ldots, \boldsymbol{A}^{n-1}\boldsymbol{b}\}, \tag{1.46}$$

where $\boldsymbol{A}^0 = \boldsymbol{I}$.

If we use an iterative procedure to obtain the approximate solution $\boldsymbol{u}_k$ to $\boldsymbol{A}\boldsymbol{u} = \boldsymbol{b}$ at kth iteration, the residual is given by

$$\boldsymbol{r}_k = \boldsymbol{b} - \boldsymbol{A}\boldsymbol{u}_k, \tag{1.47}$$

which is essentially the negative gradient $\nabla f(\boldsymbol{u}_k)$. The search direction vector in the conjugate gradient method is subsequently determined by

$$\boldsymbol{d}_{k+1} = \boldsymbol{r}_k - \frac{\boldsymbol{d}_k^T \boldsymbol{A}\boldsymbol{r}_k}{\boldsymbol{d}_k^T \boldsymbol{A}\boldsymbol{d}_k}\boldsymbol{d}_k. \tag{1.48}$$

The solution often starts with an initial guess $\boldsymbol{u}_0$ at $k = 0$, and proceeds iteratively. The above steps can compactly be written as

$$\boldsymbol{u}_{k+1} = \boldsymbol{u}_k + \alpha_k \boldsymbol{d}_k, \quad \boldsymbol{r}_{k+1} = \boldsymbol{r}_k - \alpha_k \boldsymbol{A}\boldsymbol{d}_k, \tag{1.49}$$

and

$$\boldsymbol{d}_{k+1} = \boldsymbol{r}_{k+1} + \beta_k \boldsymbol{d}_k, \tag{1.50}$$

where

$$\alpha_k = \frac{\boldsymbol{r}_k^T \boldsymbol{r}_k}{\boldsymbol{d}_k^T \boldsymbol{A}\boldsymbol{d}_k}, \quad \beta_k = \frac{\boldsymbol{r}_{k+1}^T \boldsymbol{r}_{k+1}}{\boldsymbol{r}_k^T \boldsymbol{r}_k}. \tag{1.51}$$

Iterations stop when a prescribed accuracy is reached. This can easily be programmed in any programming language, especially Matlab.

It is worth pointing out that the initial guess $\boldsymbol{r}_0$ can be any educated guess; however, $\boldsymbol{d}_0$ should be taken as $\boldsymbol{d}_0 = \boldsymbol{r}_0$, otherwise, the algorithm may not converge. In the case when $\boldsymbol{A}$ is not symmetric, we can use the generalized minimal residual (GMRES) algorithm developed by Y. Saad and M.H. Schultz in 1986. For a more comprehensive literature review, readers can refer to Press et al. [86].

1.5.5 Stochastic Gradient Descent

In many optimization problems, especially in deep learning [17, 62], the objective function or risk function to be minimized can be written in the following form:

$$E(\boldsymbol{w}) = \frac{1}{m}\sum_{i=1}^{m} f_i(\boldsymbol{x}_i, \boldsymbol{w}) = \frac{1}{m}\sum_{i=1}^{m}\big[u_i(\boldsymbol{x}_i, \boldsymbol{w}) - \bar{y}_i\big]^2, \tag{1.52}$$

where

$$f_i(\boldsymbol{x}_i, \boldsymbol{w}) = \big[u_i(\boldsymbol{x}_i, \boldsymbol{w}) - \bar{y}_i\big]^2. \tag{1.53}$$

Here, $\boldsymbol{w} = (w_1, w_2, \ldots, w_K)^T$ is a parameter vector such as the weights in a neural networks. In addition, $\bar{y}_i (i = 1, 2, \ldots, m)$ are m targets or real data (data points or data sets), while $u_i(\boldsymbol{x}_i)$ are the predicted values based on the inputs $\boldsymbol{x}_i$ by a model such as the models based on trained neural networks.

The standard gradient descent for finding new weight parameters in terms of iterative formula can be written as

$$\boldsymbol{w}^{t+1} = \boldsymbol{w}^t - \frac{\eta}{m}\sum_{i=1}^{m} \nabla f_i, \tag{1.54}$$

where $0 < \eta \leq 1$ is the learning rate or step size. Here, the gradient ∇f_i is with respect to $\boldsymbol{w}$. This requires the calculations of m gradients. When m is large and the number of iterations t is large, this can be very expensive.

In order to save computation, the true gradient can be approximated by the gradient at a single value at f_i instead of all m values. That is,

$$\boldsymbol{w}^{t+1} = \boldsymbol{w}^t - \eta_t \nabla f_i, \tag{1.55}$$

where η_t is the learning rate at iteration t, which can be varying with iterations. Though this is a crude estimate at a randomly selected point i at iteration t to the true gradient, the computation costs have dramatically reduced by a factor of $1/m$. Due to the random nature of the selection of a sample i (which can be very different at each iteration), this way of calculating gradient is called stochastic gradient. The method based on such crude estimation of gradients is called stochastic gradient descent (SGD) for minimization or stochastic gradient ascent (SGA) for maximization.

The learning rate η_t should be reduced gradually. For example, a commonly used reduction of learning rates is

$$\eta_t = \frac{1}{1 + \beta t}, \qquad t = 1, 2, \ldots, \tag{1.56}$$

where $\beta > 0$ is a hyper-parameter. Bottou [17] showed that SGD will almost surely converge if

$$\sum_t \eta_t = \infty, \qquad \sum_t \eta_t^2 < \infty. \tag{1.57}$$

The best convergence rate is $\eta_t \sim 1/t$ with the averaged residual error decreasing in the manner of $E \sim 1/t$.

It is worth pointing out that the stochastic gradient descent is not the direct descent in the true gradient sense, but the descent is in terms of average or expectation. Thus, the paths can still be zig-zag, sometimes, it may be up the gradient, not necessarily all the way down the gradient directions, but the overall computation efficiency is usually much higher than the true gradient descent for large-scale problems. Therefore, it is widely used for deep learning problems and large-scale problems.

1.5.6 Subgradient Method

All the above gradient-based methods assume implicitly that the functions are differentiable. In the case of non-differentiable functions, we have to use the subgradient method for non-differential convex functions or more generally gradient-free methods for nonlinear functions to be introduced later in this book.

For non-differentiable convex functions, there may be more than one subgradient vector at any point, and subgradient vectors $\boldsymbol{v}_k$ can be defined by

$$f(\boldsymbol{x}) - f(\boldsymbol{x}_k) \geq \boldsymbol{v}_k^T (\boldsymbol{x} - \boldsymbol{x}_k), \tag{1.58}$$

and Newton's iteration formula (1.37) can be replaced by

$$\boldsymbol{x}^{k+1} = \boldsymbol{x}^k - \alpha_k \boldsymbol{v}_k, \tag{1.59}$$

where α_k is the step size at iteration k. As the iteration formula involves the subgradient $\boldsymbol{v}_k = \partial f(\boldsymbol{x}_k)$ calculated at iteration k, the method is called the subgradient method.

It is worth pointing out that since there are many arbitrary subgradients, the subgradient calculated at $\boldsymbol{x}_k$ may not be in the desirable direction. Some choices such as choosing the larger values of the norm $\boldsymbol{v}$ can be expected.

In addition, though a constant step size $\alpha_k = \alpha$ where $0 < \alpha < 1$ can work well in many cases, it is desirable that the step size α_k should vary and be scaled when appropriate. For example, a commonly used scheme for varying step sizes is

$$\alpha_k \geq 0, \quad \sum_{k=1}^{\infty} \alpha_k^2 < \infty, \tag{1.60}$$

and

$$\lim_{k \to \infty} \alpha_k = 0. \tag{1.61}$$

The subgradient method is still used in practice, and it can be very effective in combination with the stochastic gradient method, which leads to a class of the so-called stochastic subgradient methods. Convergence can be proved and interested readers can refer to more advanced literature such as Bertsbekas et al. [14].

The limitation of the subgradient method is that it is mainly for convex functions. In case of general nonlinear, non-differentiable, non-convex functions, we can use gradient-free methods and we will introduce some of these methods in the next chapter.

Chapter 2
Nature-Inspired Algorithms

The literature of nature-inspired algorithms and swarm intelligence is expanding rapidly, here we will introduce some of the most recent and widely used nature-inspired optimization algorithms [114, 118, 131].

2.1 A Brief History of Nature-Inspired Algorithms

Most traditional algorithms such as Newton's method are deterministic algorithms in the sense that a given starting point will lead to exactly the same sequence of solutions or points [9, 13, 21, 96, 118]. On the other hand, nature-inspired algorithms are usually stochastic or non-deterministic algorithms.

The initiation of non-deterministic algorithms was in the 1940s by Alan Turing [27, 103]. In the 1960s, evolutionary strategy and genetic algorithm started to appear, which attempted to simulate the key feature of Darwinian evolution of biological systems. For example, genetic algorithm (GA) was developed by John Holland in the 1960s [49], which uses crossover, mutation and selection as basic genetic operators for algorithm operations. At about the same period, Ingo Recehberg and H.P. Schwefel developed the evolutionary strategy for constructing automatic experimenter using simple rules of mutation and selection, though crossover was not used. In around 1966, L.J. Fogel and colleagues used simulated evolution as a learning tool to study artificial intelligence, which leads to the development of evolutionary programming. All these algorithms now evolved into a much wider discipline, called evolutionary algorithms or evolutionary computation [37]. All these algorithms can also be considered as part of heuristics [54].

Then, simulated annealing was developed in 1983 by Kirpatrick et al. [59], which simulates the annealing process of metals for the optimization purpose, and the Tabu search was developed by Fred Glover in 1986 [44] that uses memory and history

X.-S. Yang, X.-S. He, *Mathematical Foundations of Nature-Inspired Algorithms*,
SpringerBriefs in Optimization, https://doi.org/10.1007/978-3-030-16936-7_2

to enhance the search efficiency. In fact, it was Fred Glover who coined the word 'metaheuristic' in his 1986 paper [44].

The major development in the context of swarm intelligence based algorithms started in the 1990s. First, Marco Dorigo developed the ant colony optimization (ACO) in his PhD work [31], and ACO uses the key characteristics of social ants to design procedure for optimization. Local interactions via pheromone and rules are used in ACO. Then, in 1995, particle swarm optimization was developed by James Kennedy and Russell C. Eberhardt, inspired by the swarming behaviour of fish and birds [57], which forms part of the key ideas of contemporary swarm intelligence [33, 34, 58]. In about 1992, Koza developed genetic programming by means of natural selection [60].

Though developed in 1997, differential evolution (DE) is not a nature-inspired algorithm; however, DE has used vectorized mutation which forms a basis for many later algorithms [95].

Another interesting development is a theory concerning no-free-lunch (NFL) theorems, proved in 1997 by D.H. Wolpert and W.G. Macready, which had much impact in the optimization and machine learning communities [107]. This basically dashed the dreams for finding the best algorithms for solving all problems effectively because NFL theorems state that all algorithms are equally effective or ineffective if measured in terms of averaged performance for *all* possible problems [53, 107]. They are essentially equivalent to any random search algorithm in the *average* sense. Then, researchers realized that the performance and efficiency in practice are not measured by averaging over all possible problems. Instead, we are more concerned with a particular class of problems in a particular discipline, and there is no need to use an algorithm to solve all possible problems. Consequently, for a finite set of problems and for a given few algorithms, empirical observations and experience suggest that some algorithms can perform better than others. For example, algorithms that can use problem-specific knowledge such as convexity can be more efficient than random search. Therefore, further research should identify the types of problems that a given algorithm can solve, or the most suitable algorithms for a given type of problems. Thus, research resumes and continues, just with a different emphasis and from different perspectives. In addition, later studies also suggested that free lunch may exist for co-evolutionary systems [108].

At the turn of this century, developments of nature-inspired algorithms became more active. In 2004, a honeybee algorithm for optimizing Internet hosting centres was developed by Sunil Nakrani and Craig Tovey [71]. In 2005, Pham et al. [85] developed the bees algorithm, and the virtual bee algorithm was developed by Xin-She Yang in 2005 [109]. About the same time, the artificial bee colony (ABC) algorithm was developed by D. Karaboga in 2005 [55]. All these algorithms are bee-based algorithms and they all use some (but different) aspects of the foraging behaviour of social bees.

Then, in late 2007 and early 2008, the firefly algorithm (FA) was developed by Xin-She Yang, inspired by the flashing behaviour of tropic firefly species [110]. The attraction mechanism, together with the variation of light intensity, was used to produce a nonlinear algorithm that can deal with multimodal optimization problems. In 2009, cuckoo search (CS) was developed by Xin-She Yang and Suash Deb, inspired by the brood parasitism of the reproduction strategies of some cuckoo species [122]. This algorithm simulates partly the complex social interactions of cuckoo–host species co-evolution. Then, in 2010, the bat algorithm (BA) was developed by Xin-She Yang, inspired by the echolocation characteristics of microbats [112]. BA uses frequency tuning in combination with the variations of loudness and pulse emission rates during foraging. All these algorithms can be considered as swarm intelligence based algorithms because they use the 'social' interactions and their biologically inspired rules [4, 12].

There are other algorithms developed in the last two decades, but they are not swarm intelligence based algorithms. For example, harmony search, developed in 2001, is a music-inspired algorithm [42], while gravitational search algorithm (GSA) is a physics-inspired algorithm [87]. In addition, flower pollination algorithm (FPA) is an algorithm inspired by the pollination features of flowering plants [116] with promising applications [89, 132]. All these algorithms are population-based algorithms, but they do not strictly belong to swarm intelligence algorithms.

A wider range of applications of nature-inspired algorithms can be found in the recent literature [117, 118, 121, 128]. In the rest of chapter, we will briefly introduce some of the most recent nature-inspired algorithms.

2.2 Genetic Algorithms

The genetic algorithm (GA) is an evolutionary algorithm and probably the most widely used. It is becoming a conventional and classic method. However, it does have fundamental genetic operators that have inspired many later algorithms so we will introduce it in detail.

The genetic algorithm (GA), developed by John Holland and his collaborators in the 1960s and 1970s, is a model or abstraction of biological evolution based on Charles Darwin's theory of natural selection. Holland [49] was the first to use crossover and recombination, together with mutation and selection, in the study of adaptive and artificial systems. These genetic operators form the essential part of the genetic algorithm as a problem-solving strategy. Since then, many variants of genetic algorithms have been developed and applied to a wide range of optimization problems, from graph colouring to pattern recognition, from discrete systems (such as the travelling salesman problem) to continuous systems (e.g., the efficient design of airfoils in aerospace engineering), and from the financial market to multiobjective engineering optimization.

There are many advantages of metaheuristic algorithms such as genetic algorithms over traditional optimization algorithms, the two most noticeable advantages are the ability to deal with complex problems, and parallelism. Genetic algorithms can deal with various types of optimization whether the objective (fitness) function is stationary or non-stationary (change with time), linear or nonlinear, continuous or discontinuous, or with random noise. As multiple offspring in a population act like independent agents, the population (or any subgroup) can explore the search space in many directions simultaneously. This feature makes it ideal to parallelize the algorithm for implementation. Different parameters and even different groups of strings can be manipulated at the same time. Such advantages also map onto the algorithms based on swarm intelligence (SI) and thus SI-based algorithms such as particle swarm optimization and firefly algorithm to be introduced later also possess such good advantages.

However, genetic algorithms also have some disadvantages. The formulation of the fitness function, the population size, the choice of the important parameters such as the rate of mutation and crossover, and the selection criteria of new populations should be carried out carefully. Any inappropriate choice will make it difficult for the algorithm to converge, or it simply produces meaningless results.

There are many variants of the genetic algorithm, and they now form a class of genetic algorithms [46, 49]. The essence of genetic algorithms involves the encoding of an objective function as arrays of bits or character strings to represent the chromosomes, the manipulation operations of strings by genetic operators, and the selection according to their fitness with the aim of finding a solution to the problem concerned. This is often done by the following procedure: (1) encoding of solutions into strings; (2) defining a fitness function and selection criterion; (3) creating a population of individuals and evaluating their fitness; (4) evolving the population by generating new solutions using crossover, mutation, fitness-proportionate reproduction; (5) selecting new solutions according to their fitness and replacing the old population by better individuals; (6) decoding the results to obtain the solution(s) to the problem.

An important issue is the formulation or choice of an appropriate fitness function that determines the selection criterion in a particular problem. For the minimization of $f(\boldsymbol{x})$ using genetic algorithms, one simple way of constructing a fitness function is to use the simplest form $F(\boldsymbol{x}) = A - f(\boldsymbol{x})$ with A being a large constant (though $A = 0$ will do), thus the objective is to maximize the fitness function. However, there are many different ways of defining a fitness function. For example, we can use the individual fitness assignment relative to the whole population

$$F(\boldsymbol{x}_i) = \frac{f(\boldsymbol{x}_i))}{\sum_{i=1}^{N} f(\boldsymbol{x}_i)}, \tag{2.1}$$

where N is the population size. The appropriate form of the fitness function will ensure that the solutions with higher fitness should be selected efficiently. Poorly defined fitness functions may result in incorrect or meaningless solutions.

Another important issue is the choice of various parameters. The crossover probability p_c is usually very high, typically in the range of 0.7–0.99. On the other hand, the mutation probability p_m is usually small (usually 0.001–0.05). If p_c is too small, then the crossover occurs sparsely, which is not efficient for evolution. If the mutation probability is too high, the diversity of the population may be too high, which makes it harder for the system to converge.

The selection criterion is also important; how to select the current population so that the best individuals with higher fitness are preserved and passed on to the next generation. That is often carried out in association with a certain elitism. The basic elitism is to select the fittest individual (in each generation) which will be carried over to the new generation without being modified by genetic operators. This ensures that the best solution is achieved more quickly.

Other issues include multiple sites for mutation and crossover. Mutation at a single site is not very efficient; mutation at multiple sites will increase the evolution efficiency. However, too many mutants will make it difficult for the system to converge, or even make the system go astray to the wrong solutions. In reality, if the mutation is too high under high selection pressure, then the whole population might go extinct. Similarly, crossover can also be carried out at multiple parts, which can increase the mixing ability of the population and increase the efficiency of crossover.

In addition, the choice of the right population size is also very important. If the population size is too small, there is not enough evolution going on, and there is a risk that the whole population may go extinct. In the real world, for a species with a small population, ecological theory suggests that there is a real danger of extinction. Even though the system carries on, there is still a danger of premature convergence. In a small population, if a significantly more fit individual appears too early, it may reproduce enough offspring to overwhelm the whole (small) population. This will eventually drive the system to a local optimum (not the global optimum). On the other hand, if the population is too large, more evaluations of the objective function are needed, which will require an extensive computing time.

2.3 Simulated Annealing

Simulated annealing (SA) is a random search technique for global optimization problems, and it mimics the annealing process in materials processing when a metal cools and freezes into a crystalline state with the minimum energy and larger crystal size so as to reduce the defects in metallic structures. The annealing process involves the careful control of temperature and cooling rate (often called annealing schedule).

The application of simulated annealing into optimization problems was pioneered by Kirkpatrick, Gelatt and Vecchi in 1983 [59]. Since then, there have been extensive studies. Unlike the gradient-based methods and other deterministic search methods which have the disadvantage of becoming trapped in local minima, the main advantage of simulated annealing is its ability to avoid being trapped in local minima. In fact, it has been proved that simulated annealing will converge to its global optimality if enough randomness is used in combination with very slow cooling.

Metaphorically speaking, the iterations in SA are equivalent to dropping some bouncing balls over a landscape. As the balls bounce and lose energy, they will settle down to some local minima. If the balls are allowed to bounce enough times and lose energy slowly enough, some of the balls will eventually fall into the lowest global locations, hence the global minimum will be reached.

The basic idea of the simulated annealing algorithm is to use random search which not only accepts changes that improve the objective function, but also keeps some changes that are not ideal. In a minimization problem, for example, any better moves or changes that decrease the cost (or the value) of the objective function f will be accepted; however, some changes that increase f will also be accepted with a probability p. This probability p, also called the transition probability, is determined by

$$p = \exp[-\frac{\delta E}{k_B T}], \tag{2.2}$$

where k_B is the Boltzmann constant, and T is the temperature for controlling the annealing process. δE is the change of the energy level. This transition probability is based on the Boltzmann distribution in physics. The simplest way to link δE with the change of the objective function δf is to use $\delta E = \gamma \delta f$, where γ is a real constant. For simplicity without losing generality, we can use $k_B = 1$ and $\gamma = 1$. Thus, the probability p simply becomes

$$p(\delta f, T) = e^{-\frac{\delta f}{T}}. \tag{2.3}$$

Whether or not to accept a change, we usually use a random number r (drawn from a uniform distribution in [0,1]) as a threshold. Thus, if $p > r$ or

$$p = e^{-\frac{\delta f}{T}} > r, \tag{2.4}$$

it is accepted.

Here the choice of the right temperature is crucially important. For a given change δf, if T is too high ($T \to \infty$), then $p \to 1$, which means almost all changes will be accepted. If T is too low ($T \to 0$), then any $\delta f > 0$ (worse solution) will rarely be accepted as $p \to 0$ and thus the diversity of the solution is limited, but any improvement δf will almost always be accepted. In fact, the special case $T \to 0$ corresponds to the gradient-based method because only better solutions are accepted, and the system is essentially climbing up or descending a hill. Therefore, if T is too high, the system is at a high energy state on the topological landscape, and the minima are not easily reached. If T is too low, the system may be trapped in a local minimum (not necessarily the global minimum), and there is not enough energy for the system to jump out of the local minimum to explore other potential global minima. So a proper, initial temperature should be calculated.

Another important issue is how to control the cooling process so that the system cools down gradually from a higher temperature to ultimately freeze to a global minimum state. There are many ways to control the cooling rate or the decrease in temperature.

Two commonly used cooling schedules are: linear and geometric cooling. For a linear cooling process, we have

$$T = T_0 - \beta t,$$

(or $T \to T - \delta T$ with a temperature increment δT). Here, T_0 is the initial temperature, and t is the pseudotime for iterations. β is the cooling rate, and it should be chosen in such a way that $T \to 0$ when $t \to t_f$ (maximum number of iterations), which usually gives $\beta = T_0/t_f$.

The geometric cooling essentially decreases the temperature by a cooling factor $0 < \alpha < 1$ so that T is replaced by αT or

$$T(t) = T_0 \alpha^t, \quad t = 1, 2, \ldots, t_f. \tag{2.5}$$

The advantage of the second method is that $T \to 0$ when $t \to \infty$, and thus there is no need to specify the maximum number of iterations t_f. For this reason, we will use this geometric cooling schedule. The cooling process should be slow enough to allow the system to stabilize easily. In practice, $\alpha = 0.7$–0.99 is commonly used.

In addition, for a given temperature, multiple evaluations of the objective function are needed. If there are too few evaluations, there is a danger that the system will not stabilize and subsequently will not converge to its global optimality. If there are too many evaluations, it is time-consuming, and the system will usually converge too slowly as the number of iterations to achieve stability may be exponential to the problem size.

Therefore, there is a balance between the number of evaluations and solution quality. We can either do many evaluations at a few temperature levels or do few evaluations at many temperature levels. There are two major ways to set the number of iterations: fixed or varied. The first uses a fixed number of iterations at each temperature, while the second is designed to increase the number of iterations at lower temperatures so that the local minima can be fully explored.

In order to find a suitable starting temperature T_0, we can use any information about the objective function. If we know the maximum change $\max(\delta f)$ of the objective function, we can use it to estimate an initial temperature T_0 for a given probability p_0. That is,

$$T_0 \approx -\frac{\max(\delta f)}{\ln p_0}. \tag{2.6}$$

If we do not know the possible maximum change of the objective function, we can use a heuristic approach. We can start evaluations from a very high temperature (so that almost all changes are accepted) and reduce the temperature quickly until about 50 or 60% of the worse moves are accepted, and then use this temperature as the new initial temperature T_0 for proper and relatively slow cooling.

For the final temperature, in theory it should be zero so that no worse move can be accepted. However, if $T_f \to 0$, more unnecessary evaluations are needed. In practice, we simply choose a very small value, say, $T_f = 10^{-10}$–10^{-5}, depending on the required quality of the solutions and time constraints.

The implementation of simulated annealing to optimize the banana function can be demonstrated using a Matlab/Octave program. We have used the initial temperature $T_0 = 1.0$, the final temperature $T_f = 10^{-10}$ and a geometric cooling schedule with $\alpha = 0.9$ [118]. A demo code in Matlab can be found at the Mathworks.[1]

2.4 Ant Colony Optimization

Ants are social insects that live together in well-organized colonies with a population size ranging from about 2 to 25 million. Ants communicate with each other and interact with their environment in a swarm using local rules and scent chemicals or pheromone. There is no centralized control. Such a complex system with local interactions can self-organize with emerging behaviour, leading to some form of social intelligence [16].

Based on these characteristics, the ant colony optimization (ACO) was developed by Marco Dorigo in 1992 [31], and ACO attempts to mimic the foraging behaviour of social ants in a colony. Pheromone is deposited by each agent, and such chemical will also evaporate. The model for pheromone deposition and evaporation may vary slightly, depend on the variants of ACO. However, in most cases, incremental deposition and exponential decay are used in the literature.

From the implementation point of view, for example, a solution in a network optimization problem can be a path or route. Ants will explore the network paths and deposit pheromone when it moves. The quality of a solution is related to the pheromone concentration on the path. At the same time, pheromone will evaporate as (pseudo)time increases. At a junction with multiple routes, the probability of choosing a particular route is determined by a decision criterion, depending on the normalized concentration of the route, and relative fitness of this route, comparing with all others. For example, in most studies, the probability p_{ij} of choosing a route from node i to node j can be calculated by

$$p_{ij} = \frac{\phi_{ij}^{\alpha} d_{ij}^{\beta}}{\sum_{i,j}^{n} \phi_{ij}^{\alpha} d_{ij}^{\beta}}, \tag{2.7}$$

where $\alpha, \beta > 0$ are the so-called influence parameters, and ϕ_{ij} is the pheromone concentration on the route between i and j. In addition, d_{ij} is the desirability of the route (for example, the distance of the overall path). In the simplest case

[1] http://www.mathworks.co.uk/matlabcentral/fileexchange/29739-simulated-annealing-for-constrained-optimization.

when $\alpha = \beta = 1$, the choice probability is simply proportional to the pheromone concentration.

It is worth pointing out that ACO is a mixed of procedure and some simple equations such as pheromone deposition and evaporation as well as the path selection probability. ACO has been applied to many applications from scheduling to routing problems [31].

2.5 Differential Evolution

Differential evolution (DE) was developed by R. Storn and K. Price in 1996 and 1997 [95]. It is a vector-based algorithm, which has some similarity to pattern search and genetic algorithms due to its use of crossover and mutation. DE is a stochastic search algorithm with self-organizing tendency and does not use the information of derivatives. Thus, it is a population-based, derivative-free method. In addition, DE uses real number as solution strings; thus, no encoding and decoding is needed.

For a D-dimensional optimization problem with D parameters, a population of n solution vectors are initially generated, we have $\boldsymbol{x}_i$ where $i = 1, 2, \ldots, n$. For each solution $\boldsymbol{x}_i$ at any generation t, we use the conventional notation as

$$\boldsymbol{x}_i^t = (x_{1,i}^t, x_{2,i}^t, \ldots, x_{D,i}^t), \tag{2.8}$$

which consists of D components in the D-dimensional space. This vector can be considered as the chromosomes or genomes.

Differential evolution consists of three main steps: mutation, crossover and selection.

Mutation is carried out by the mutation scheme. For each vector $\boldsymbol{x}_i$ at any time or generation t, we first randomly choose three distinct vectors $\boldsymbol{x}_p$, $\boldsymbol{x}_q$ and $\boldsymbol{x}_r$ at t, and then generate a so-called donor vector by the mutation scheme

$$\boldsymbol{v}_i^{t+1} = \boldsymbol{x}_p^t + F(\boldsymbol{x}_q^t - \boldsymbol{x}_r^t), \tag{2.9}$$

where $F \in [0, 2]$ is a parameter, often referred to as the differential weight. This requires that the minimum number of population size is $n \geq 4$. In principle, $F \in [0, 2]$, but in practice, a scheme with $F \in [0, 1]$ is more efficient and stable. In fact, almost all the studies in the literature use $F \in (0, 1)$.

Crossover is controlled by a crossover parameter $C_r \in [0, 1]$, controlling the rate or probability for crossover. The actual crossover can be carried out in two ways: binomial and exponential. The binomial scheme performs crossover on each of the D components or variables/parameters. By generating a uniformly distributed random number $r_i \in [0, 1]$, the jth component of $\boldsymbol{v}_i$ is manipulated as

$$\boldsymbol{u}_{j,i}^{t+1} = \begin{cases} \boldsymbol{v}_{j,i} & \text{if } r_i \leq C_r, \\ \boldsymbol{x}_{j,i}^t & \text{otherwise,} \end{cases} \qquad j = 1, 2, \ldots, D. \tag{2.10}$$

This way, each component can be decided randomly whether or not to exchange with the counterpart of the donor vector.

In the exponential scheme, a segment of the donor vector is selected and this segment starts with a random integer k with a random length L which can include many components. Mathematically, this is to choose $k \in [0, D-1]$ and $L \in [1, D]$ randomly, and we have

$$\boldsymbol{u}_{j,i}^{t+1} = \begin{cases} \boldsymbol{v}_{j,i}^t & \text{for } j = k, \ldots, k - L + 1 \in [1, D], \\ \boldsymbol{x}_{j,i}^t & \text{otherwise.} \end{cases} \tag{2.11}$$

As the binomial is simpler to implement, we will use the binomial crossover in our discussions here.

Selection is essentially the same as that used in genetic algorithms. It is to select the most fittest, that is, the minimum objective value for a minimization problem. Therefore, we have

$$\boldsymbol{x}_i^{t+1} = \begin{cases} \boldsymbol{u}_i^{t+1} & \text{if } f(\boldsymbol{u}_i^{t+1}) \leq f(\boldsymbol{x}_i^t), \\ \boldsymbol{x}_i^t & \text{otherwise.} \end{cases} \tag{2.12}$$

It is worth pointing out here that the use of $\boldsymbol{v}_i^{t+1} \neq \boldsymbol{x}_i^t$ may increase the evolutionary or exploratory efficiency. The overall search efficiency is controlled by two parameters: the differential weight F and the crossover probability C_r.

Most studies have focused on the choice of F, C_r and n as well as the modifications of (2.9). In fact, when generating mutation vectors, we can use many different ways of formulating (2.9), and this leads to various schemes with the naming convention: DE/x/y/z, where x is the mutation scheme (rand or best), y is the number of difference vectors and z is the crossover scheme (binomial or exponential). So DE/Rand/1/* means the basic DE scheme using random mutation, one difference vector with either a binomial or exponential crossover scheme.

The basic DE/Rand/1/Bin scheme is given in (2.9). That is,

$$\boldsymbol{v}_i^{t+1} = \boldsymbol{x}_p^t + F(\boldsymbol{x}_q^t - \boldsymbol{x}_r^t). \tag{2.13}$$

If we replace the $\boldsymbol{x}_p^t$ by the current best $\boldsymbol{x}_{\text{best}}$ found so far, we have the so-called DE/Best/1/Bin scheme

$$\boldsymbol{v}_i^{t+1} = \boldsymbol{x}_{\text{best}}^t + F(\boldsymbol{x}_q^t - \boldsymbol{x}_r^t). \tag{2.14}$$

There is no reason that why we should not use more than three distinct vectors. For example, if we use four different vectors plus the current best, we have the DE/Best/2/Bin scheme

$$v_i^{t+1} = x_{\text{best}}^t + F(x_{k_1}^t + x_{k_2}^t - x_{k_3}^t - x_{k_4}^t). \tag{2.15}$$

Furthermore, if we use five different vectors, we have the DE/Rand/2/Bin scheme

$$v_i^{t+1} = x_{k_1}^t + F_1(x_{k_2}^t - x_{k_3}^t) + F_2(x_{k_4}^t - x_{k_5}^t), \tag{2.16}$$

where F_1 and F_2 are differential weights in [0, 1]. Obviously, for simplicity, we can also take $F_1 = F_2 = F$. Following the similar strategy, we can design various schemes. For example, the above variants can be written in a generalized form

$$v_i^{t+1} = x_{k_1}^t + \sum_{s=1}^{m} F_s \cdot (x_{k_2(s)}^t - x_{k_3(s)}^t), \tag{2.17}$$

where $m = 1, 2, 3, \ldots$ and $F_s (s = 1, \ldots, m)$ are the scale factors. The number of vectors involved on the right-hand side is $2m + 1$. In the above variants, $m = 1$ and $m = 2$ are used.

2.6 Particle Swarm Optimization

Many swarms in nature such as fish and birds can have higher-level behaviour, but they all obey simple rules. For example, a swarm of birds such as starlings simply follow three basic rules: each bird flies according to the flight velocities of their neighbour birds (usually about seven adjacent birds), whiling keep a certain separation distance. Birds on the edge of the swarm tend to fly into the centre of the swarm (so as to avoid being eaten by potential predators such as eagles). In addition, birds tend to fly to search for food or shelters; thus, a short memory is used. Based on such swarming characteristics, particle swarm optimization (PSO) was developed by Kennedy and Eberhart in 1995, which uses equations to simulate the swarming characteristics of birds and fish [57].

For the ease of discussions below, let us use x_i and v_i to denote the position (solution) and velocity, respectively, of a particle or agent i. In PSO, there are n particles as a population, thus $i = 1, 2, \ldots, n$. There are two equations for updating positions and velocities of particles, and they can be written as follows:

$$v_i^{t+1} = v_i^t + \alpha\epsilon_1[g^* - x_i^t] + \beta\epsilon_2[x_i^* - x_i^t], \tag{2.18}$$

$$x_i^{t+1} = x_i^t + v_i^{t+1}\Delta t, \tag{2.19}$$

where ϵ_1 and ϵ_2 are two uniformly distributed random numbers in [0,1]. The learning parameters α and β are usually in the range of [0,2]. In the above equations, g^* is the best solution found so far by all the particles in the population, and each particle has an individual best solution x_i^* by itself during the entire past iteration history.

It is worth pointing out that $\Delta t = 1$ should be used because iterations in algorithms are discrete with a step counter $t \leftarrow t + 1$. Thus, there is no need to consider units and Δt in all algorithms discussed in this book.

It is clearly seen that the above algorithmic equations are linear in the sense that both equations only depend on $\boldsymbol{x}_i$ and $\boldsymbol{v}_i$ linearly. PSO has been applied in many applications, and it has been extended to solve multiobjective optimization problems [33, 58]. However, there are some drawbacks because PSO can often have the so-called premature convergence when the population loses diversity and thus gets stuck locally. Consequently, there are more than 20 different variants to try to remedy this with various degrees of improvements.

There are many variants which extend the standard PSO algorithm, and the most noticeable improvement is probably to use inertia function $\theta(t)$ so that $\boldsymbol{v}_i^t$ is replaced by $\theta(t)\boldsymbol{v}_i^t$

$$\boldsymbol{v}_i^{t+1} = \theta \boldsymbol{v}_i^t + \alpha \epsilon_1 [\boldsymbol{g}^* - \boldsymbol{x}_i^t] + \beta \epsilon_2 [\boldsymbol{x}_i^{*(k)} - \boldsymbol{x}_i^t], \tag{2.20}$$

where θ takes the values between 0 and 1 in theory. In the simplest case, the inertia function can be taken as a constant, typically $\theta \approx 0.5$–0.9. This is equivalent to introducing a virtual mass to stabilize the motion of the particles, and thus the algorithm can usually be expected to converge more quickly.

A simplified version which could accelerate the convergence of the algorithm is to use the global best only. The so-called accelerated particle swarm optimization (APSO) was developed by Xin-She Yang in 2008 and then has been developed further in recent years. Thus, in APSO [110], the velocity vector is generated by a simpler formula

$$\boldsymbol{v}_i^{t+1} = \boldsymbol{v}_i^t + \beta(\boldsymbol{g}^* - \boldsymbol{x}_i^t) + \alpha(\boldsymbol{\epsilon} - 1/2), \tag{2.21}$$

where $\boldsymbol{\epsilon}$ is a random variable with values from 0 to 1. Here the shift 1/2 is purely for convenience. We can also use a standard normal distribution $\alpha \boldsymbol{\epsilon}_t$, where $\boldsymbol{\epsilon}_t$ is drawn from $N(0, 1)$ to replace the second term. Now we have

$$\boldsymbol{v}_i^{t+1} = \boldsymbol{v}_i^t + \beta(\boldsymbol{g}^* - \boldsymbol{x}_i^t) + \alpha \boldsymbol{\epsilon}_t, \tag{2.22}$$

where $\boldsymbol{\epsilon}_t$ can be drawn from a Gaussian distribution or any other suitable distributions. Here, α is a scaling factor that controls the step size or the strength of randomness, while β is a parameter that controls the movement of particles.

The update of the position is simply

$$\boldsymbol{x}_i^{t+1} = \boldsymbol{x}_i^t + \boldsymbol{v}_i^{t+1}. \tag{2.23}$$

In order to simplify the formulation even further, we can also write the update of the location in a single step

$$\boldsymbol{x}_i^{t+1} = (1 - \beta)\boldsymbol{x}_i^t + \beta \boldsymbol{g}^* + \alpha \boldsymbol{\epsilon}_t. \tag{2.24}$$

The typical values for this accelerated PSO are $\alpha \approx 0.1$–0.4 and $\beta \approx 0.1$–0.7, though $\alpha \approx 0.2$ and $\beta \approx 0.5$ can be taken as the initial values for most unimodal objective functions. It is worth pointing out that the parameters α and β should in general be related to the scales of the independent variables $\boldsymbol{x}_i$ and the search domain. Surprisingly, this simplified APSO can have global convergence under appropriate conditions.

A further improvement to the accelerated PSO is to reduce the randomness as iterations proceed. This means that we can use a monotonically decreasing function such as

$$\alpha = \alpha_0 e^{-\gamma t}, \tag{2.25}$$

or

$$\alpha = \alpha_0 \gamma^t, \quad (0 < \gamma < 1), \tag{2.26}$$

where $\alpha_0 \approx 0.5$–1 is the initial value of the randomness parameter. Here t is the number of iterations or time steps. $0 < \gamma < 1$ is a control parameter. For example, in most implementations, we can use $\gamma = 0.9$–0.99. Obviously, other non-increasing function forms $\alpha(t)$ can also be used. In addition, these parameters should be fine-tuned to suit your optimization problems of interest, and such parameter tuning is essential for all evolutionary algorithms.

A demo implementation for APSO can be found at the MathWorks website.[2]

2.7 Bees-Inspired Algorithms

Bees such as honeybees live a colony and there are many subspecies of bees. Honeybees have three categories, including worker bees, queens and drones. The division of labour among bees is interesting. For example, worker bees forage, clean hive and defence the colony, and they have to collect and store honey. Honeybees communicate by pheromone and 'waggle dance' as well as other local interactions, depending on species [64]. Based on the foraging and social interactions of honeybees, researchers have developed various forms and variants of bees-inspired algorithms.

The first use of bees-inspired algorithms was probably by S. Nakrani and C.A. Tovey in 2004 to study web-hosting servers [71], while slightly later in 2004 and early 2005, Yang used the virtual bee algorithm to solve optimization problems [109]. At around the same time, D. Karaboga used the artificial bee colony (ABC) algorithm to carry out numerical optimization [55]. In addition, Pham et al.

[2]http://www.mathworks.co.uk/matlabcentral/fileexchange/29725-accelerated-particle-swarm-optimization.

used bees algorithm to solve continuous optimization and function optimization problems [85].

For example, in ABC, the bees are divided into three groups: forager bees, onlooker bees and scouts. For each food source, there is one forager bee who shares information with onlooker bees after returning to the colony from foraging, and the number of forager bees is equal to the number of food sources. Scout bees do random flight to explore, while a forager at a scarce food source may have to be forced to become a scout bee. The generation of a new solution $v_{i,k}$ is done by

$$v_{i,k} = x_{i,k} + \phi(x_{i,k} - x_{j,k}), \tag{2.27}$$

which is updated for each dimension $k = 1, 2, \ldots, D$ for different solutions (e.g., i and j) in a population of n bees ($i, j = 1, 2, \ldots, n$). Here, ϕ is a random number in [-1,1]. A food source is chosen by a roulette-based probability criterion, while a scout bee uses a Monte Carlo style randomization between the lower bound (L) and the upper bound (U)

$$x_{i,k} = L_k + r(U_k - L_k), \tag{2.28}$$

where $k = 1, 2, \ldots, D$ and r is a uniformly distributed random number in [0,1].

Bees-inspired algorithms have been applied in many applications with diverse characteristics and variants [55, 85].

2.8 Bat Algorithm

Bats are the only mammals with wings, and it is estimated that there are about 1000 different bat species. Their sizes can range from tiny bumblebee bats to giant bats. Most bat species use echolocation to a certain degree, though microbats extensively use echolocation for foraging and navigation. Microbats emit a series of loud, ultrasonic sound pulse and listen their echoes to 'see' their surroundings. The pulse properties vary and correlate with their hunting strategies. Depending on the species, pulse emission rates will increase when homing for prey with frequency-modulated short pulses (thus varying wavelengths to increase the detection resolution). Each pulse may last about 5–20 ms with a frequency range of 25–150 kHz, and the spatial resolution can be as small as a few millimetres, comparable to the size of insects they hunt [4].

Bat algorithm (BA), developed by Xin-She Yang in 2010, uses some characteristics of frequency tuning and echolocation of microbats [112, 115]. It also uses the variations of pulse emission rate r and loudness A to control exploration and exploitation. In the bat algorithm, main algorithmic equations for position $\boldsymbol{x}_i$ and velocity $\boldsymbol{v}_i$ for bat i are

$$f_i = f_{\min} + (f_{\max} - f_{\min})\beta, \tag{2.29}$$

$$v_i^t = v_i^{t-1} + (x_i^{t-1} - x_*) f_i, \tag{2.30}$$

$$x_i^t = x_i^{t-1} + v_i^t \Delta t, \tag{2.31}$$

where $\beta \in [0, 1]$ is a random vector drawn from a uniform distribution so that the frequency can vary from $f_{\min}$ to $f_{\max}$. Here, x_* is the current best solution found so far by all the virtual bats. As pointed out earlier, $\Delta t = 1$ is used for iterative, discrete algorithms.

From the above equations, we can see that both equations are linear in terms of x_i and v_i. But the control of exploration and exploitation is carried out by the variations of loudness $A(t)$ from a high value to a lower value and the emission rate r from a lower value to a higher value. That is,

$$A_i^{t+1} = \alpha A_i^t, \quad r_i^{t+1} = r_i^0 (1 - e^{-\gamma t}), \tag{2.32}$$

where $0 < \alpha < 1$ and $\gamma > 0$ are two parameters. As a result, the actual algorithm can have a weak nonlinearity. Consequently, BA can have a faster convergence rate in comparison with PSO. BA has been extended to multiobjective optimization, chaotic bat algorithm, and hybrid versions with a diverse range of applications [11, 40, 115, 118]. Its global convergence has been proved theoretically by Chen et al. in 2018 [24].

A demo code of the bat algorithm can be found at the MathWorks website.[3]

In the BA, frequency tuning essentially acts as mutation, while the selection pressure is relatively constant via the use of the current best solution x_* found so far. There is no explicit crossover; however, mutation varies due to the variations of loudness and pulse emission. In addition, the variations of loudness and pulse emission rates also provide an auto-zooming ability so that exploitation becomes intensive as the search moves are approaching the global optimality.

The bat algorithm and its variants have been applied to solve various types of optimization problems. For a more detailed literature, readers can refer to Yang and He [127].

2.9 Firefly Algorithm

There are about 2000 species of fireflies and most species produce short, rhythmic flashes by bioluminescence. Each species can have different flashing patterns and rhythms, and one of the main functions of such flashing light acts as a signalling system to communicate with other fireflies [63]. As light intensity in the night sky decreases as the distance from the flashing source increases, the range of visibility can be typically a few hundred metres, depending on weather conditions. The

[3] http://www.mathworks.co.uk/matlabcentral/fileexchange/37582-bat-algorithm--demo-.

attractiveness of a firefly is usually linked to the brightness of its flashes and the timing accuracy of its flashing patterns.

Based on the above characteristics, Xin-She Yang developed in 2008 the firefly algorithm (FA) [110, 111]. FA uses a nonlinear system by combining the exponential decay of light absorption and inverse-square law of light variation with distance. In the FA, the main algorithmic equation for the position $\boldsymbol{x}_i$ (as a solution vector to a problem) is

$$\boldsymbol{x}_i^{t+1} = \boldsymbol{x}_i^t + \beta_0 e^{-\gamma r_{ij}^2} (\boldsymbol{x}_j^t - \boldsymbol{x}_i^t) + \alpha\, \boldsymbol{\epsilon}_i^t, \tag{2.33}$$

where α is a scaling factor controlling the step sizes of the random walks, while γ is a scale-dependent parameter controlling the visibility of the fireflies (and thus search modes). In addition, β_0 is the attractiveness constant when the distance between two fireflies is zero (i.e., $r_{ij} = 0$). This system is a nonlinear system, which may lead to rich characteristics in terms of algorithmic behaviour.

Since the brightness of a firefly is associated with the objective landscape with its position as the indicator, the attractiveness of a firefly seen by others, depending on their relative positions and relative brightness. Thus, the beauty is in the eye of the beholder. Consequently, a pair comparison is needed for comparing all fireflies. The main steps of FA can be summarized as the pseudocode in Algorithm 1.

It is worth pointing out that α is a parameter controlling the strength of the randomness or perturbations in FA. The randomness should be gradually reduced to speed up the overall convergence. Therefore, we can use

$$\alpha = \alpha_0 \delta^t, \tag{2.34}$$

```
Initialize all the parameters α, β, γ, n;
Initialize a population of n fireflies;
Determine the light intensity/fitness at x_i by f(x_i);
while t < MaxGeneration do
    for All fireflies (i = 1 : n) do
        for All other fireflies (j = 1 : n) with i ≠ j (inner loop) do
            if Firefly j is better/brighter than i then
                Move firefly i towards j using Equation (2.33);
            end
        end
        Evaluate the new solution;
        Accept the new solution if better;
    end
    Rank and update the best solution found;
    Update iteration counter t ← t + 1;
    Reduce α (randomness strength) by a factor;
end
```

Algorithm 1: Firefly Algorithm

where α_0 is the initial value and $0 < \delta < 1$ is a reduction factor. In most cases, we can use $\delta = 0.9$–0.99, depending on the type of problems and the desired quality of solutions.

If we look at Equation (2.33) closely, we can see that γ is an important scaling parameter. At one extreme, we can set $\gamma = 0$, which means that there is no exponential decay and thus the visibility is very high (all fireflies can see each other). At the other extreme, when $\gamma \gg 1$, then the visibility range is very short. Fireflies are essentially flying in a dense fog and they cannot see each other. Thus, each firefly flies independently and randomly. Therefore, a good value of γ should be linked to the scale or limits of the design variables so that the fireflies within a range are visible to each other. This range is determined by

$$L = \frac{1}{\sqrt{\gamma}}, \tag{2.35}$$

where L the typical size of the search domain or the radius of a typical mode shape in the objective landscape. If there is no prior knowledge about its possible scale, we can start with $\gamma = 1$ for most problems.

In fact, since FA is a nonlinear system, it has the ability to automatically subdivide the whole swarm into multiple subswarms. This is because short-distance attraction is stronger than long-distance attraction, and the division of swarm is related to the mean range of attractiveness variations. After division into multi-swarms, each subswarm can potentially swarm around a local mode. Consequently, FA is naturally suitable for multimodal optimization problems. Furthermore, there is no explicit use of the best solution $\boldsymbol{g}^*$; thus, selection is through the comparison of relative brightness according to the rule of 'beauty is in the eye of the beholder'.

It is worth pointing out that FA has some significant differences from PSO. First, FA is nonlinear, while PSO is linear. Secondly, FA has an ability of multi-swarming, while PSO cannot. Thirdly, PSO uses velocities (and thus have some drawbacks), while FA does not use velocities. Finally, FA has some scaling control by using γ, while PSO has no scaling control. All these differences enable FA to search the design spaces more effectively for multimodal objective landscapes.

FA has been applied to a diverse range of real-world applications and has been extended to multiobjective optimization and hybridization with other algorithms [36, 117, 118, 133]. A simple demonstration code of the firefly algorithm can be found at the MathWorks website.[4]

[4] http://www.mathworks.co.uk/matlabcentral/fileexchange/29693-firefly-algorithm.

2.10 Cuckoo Search

In the natural world, among 141 cuckoo species, 59 species engage in the so-called obligate brood parasitism [29]. These cuckoo species do not build their own nests and they lay eggs in the nests of host birds such as warblers. Sometimes, host birds can spot the alien eggs laid by cuckoos and thus can get rid of the eggs or abandon the nest by flying away to build a new nest in a new location so as to reduce the possibility of raising an alien cuckoo chick. The eggs of cuckoos can be sufficiently similar to eggs of host birds in terms the size, colour and texture so as to increase the survival probability of cuckoo eggs. In reality, about 1/5 to 1/4 of eggs laid by cuckoos will be discovered and abandoned by hosts. In fact, there is an arms race between cuckoo species and host species, forming an interesting cuckoo–host species co-evolution system.

Based on the above characteristics, Xin-She Yang and Suash Deb developed in 2009 the cuckoo search (CS) algorithm [122, 123]. CS uses a combination of both local and global search capabilities, controlled by a discovery probability p_a. There are two algorithmic equations in CS, and one equation is

$$\boldsymbol{x}_i^{t+1} = \boldsymbol{x}_i^t + \alpha s \otimes H(p_a - \epsilon) \otimes (\boldsymbol{x}_j^t - \boldsymbol{x}_k^t), \tag{2.36}$$

where $\boldsymbol{x}_j^t$ and $\boldsymbol{x}_k^t$ are two different solutions selected randomly by random permutation, $H(u)$ is a Heaviside function, ϵ is a random number drawn from a uniform distribution and s is the step size. This step is primarily local, though it can become global search if s is large enough. However, the main global search mechanism is realized by the other equation with Lévy flights:

$$\boldsymbol{x}_i^{t+1} = \boldsymbol{x}_i^t + \alpha L(s, \lambda), \tag{2.37}$$

where the Lévy flights are simulated (or drawn random numbers) by drawing random numbers from a Lévy distribution

$$L(s, \lambda) \sim \frac{\lambda \Gamma(\lambda) \sin(\pi \lambda / 2)}{\pi} \frac{1}{s^{1+\lambda}}, \quad (s \gg 0). \tag{2.38}$$

Here $\alpha > 0$ is the step size scaling factor. It is worth pointing out that we use '$\sim$' here to highlight the fact that the steps are drawn from the distribution on the right-hand side as a sampling technique.

A simple demonstration code of the cuckoo search is provided, and the code can be found at the MathWorks website.[5]

By looking at the equations in CS carefully, we can clearly see that CS is a weakly nonlinear system due to the Heaviside function, discovery probability and Lévy flights. There is no explicit use of global best $\boldsymbol{g}^*$, but selection of the best solutions

[5]http://www.mathworks.co.uk/matlabcentral/fileexchange/29809-cuckoo-search--cs--algorithm.

is by ranking and elitism where the current best is passed onto the next generation. In addition, the use of Lévy flights can enhance the search capability because a fraction of steps generated by Lévy flights are larger than those used in Gaussian. Thus, the search steps in CS are heavy-tailed [79, 88]. Consequently, CS can be very effective for nonlinear optimization problems and multiobjective optimization [41, 68, 117, 124, 124, 135]. A relatively comprehensive literature review of cuckoo search can be found in [117, 125].

CS has two distinct advantages over other algorithms such as GA and SA, and these advantages are: efficient random walks and balanced mixing. Since Lévy flights are usually far more efficient than any other random-walk-based randomization techniques, CS can be very efficient in global search. In fact, recent studies show that CS can have guaranteed global convergence. In addition, the similarity between eggs can produce better new solutions, which is essentially fitness-proportional generation with a good mixing ability. In other words, CS has varying mutation realized by Lévy flights, and the fitness-proportional generation of new solutions based on similarity provides a subtle form of crossover. In addition, selection is carried out by using p_a where the good solutions are passed onto next generations, while not so good solutions are replaced by new solutions. Furthermore, simulations also show that CS can have auto-zooming ability in the sense that new solutions can automatically zoom into the region where the promising global optimality is located [118].

In essence, CS has strong mutation at both local and global scales, while good mixing is carried out by using solution similarity, which also plays the role of equivalent crossover. Selection is done by elitism, that is, a good fraction of solutions will be passed onto the next generation. Without the explicit use of $\boldsymbol{g}_*$, CS may also overcome the premature convergence drawback as observed in particle swarm optimization.

2.11 Flower Pollination Algorithm

Flower pollination algorithm (FPA) is a population-based algorithm, developed by Xin-She Yang and his collaborators, based on the inspiration from the pollination characteristics of flowering plants [118, 132]. FPA intends to mimic some key characteristics of biotic and abiotic pollination as well as co-evolutionary flower constancy between certain flower species and some pollinator species such as insects and animals [3, 45, 106].

In essence, there are two main equations for this algorithm, and the global search is carried out by

$$\boldsymbol{x}_i^{t+1} = \boldsymbol{x}_i^t + \gamma L(\lambda)(\boldsymbol{g}_* - \boldsymbol{x}_i^t), \tag{2.39}$$

where γ is a scaling parameter, and $L(\lambda)$ is the random number vector drawn from a Lévy distribution governed by the exponent λ, in the same form given in (2.38).

Here $\boldsymbol{g}_*$ is the best solution found so far, which acts as a selection mechanism. The current solution $\boldsymbol{x}_i^t$ is modified by varying step sizes because Lévy flights can have a fraction of large step sizes in addition to many small steps.

The local search is carried out by

$$\boldsymbol{x}_i^{t+1} = \boldsymbol{x}_i^t + U(\boldsymbol{x}_j^t - \boldsymbol{x}_k^t), \tag{2.40}$$

which mimics local pollination and flower constancy. Here, U is a uniformly distributed random number. Furthermore, $\boldsymbol{x}_j^t$ and $\boldsymbol{x}_k^t$ are solutions representing pollen from different flower patches.

The equations are linear in terms of solutions $\boldsymbol{x}_i^t$, $\boldsymbol{x}_j^t$ and $\boldsymbol{x}_k^t$, but there is a switch probability p to activate which pollination activities (global or local). As a result, the system becomes somehow quasi-linear. The randomization is achieved by three components: Lévy flights, a uniform distribution and a switch probability. As a result, FPA can typically have a higher explorative ability. At the same time, the local branch provides a mechanism to remain a strong exploitation ability. Theoretical analysis using Markov chain theory has shown that FPA can have guaranteed global convergence under the right conditions [48]. FPA has been applied to solve many optimization problems such as solar photovoltaic parameter estimation, economic and emission dispatch, and EEG-based identification [1, 2, 10, 89]. In addition, FPA has been extended to multiobjective optimization [132]. For a comprehensive review of applications, readers can refer to [5].

A demo code of the basic flower pollination algorithm can be downloaded from the MathWorks website.[6]

2.12 Other Algorithms

As we mentioned earlier, the literature is expanding and more nature-inspired algorithms are being developed by researchers, but we will not introduce more algorithms here. Instead, we will focus on summarizing the key characteristics of this class of algorithms and other population-based algorithms so as to gain a deeper understanding of these algorithms.

[6] http://www.mathworks.co.uk/matlabcentral/fileexchange/45112.

Chapter 3
Mathematical Foundations

Before we proceed to analyse any nature-inspired algorithms from at least ten different perspectives, let us review the mathematical fundamentals concerning convergence, stability and probability distributions.

3.1 Convergence Analysis

Convergence analysis is a rigorous tool for analysing how fast or slow an iterative process is, which is usually expressed as a rate of convergence towards a desired solution.

3.1.1 Rate of Convergence

Almost all algorithms are iterative and the solutions found during iterations form a sequence of $s_0, s_1, s_2, \ldots, s_n, \ldots$, that is, $s_n (n = 0, 1, 2, \ldots)$. If the sequence s_n converges towards a fixed solution (a fixed point R), we have

$$\lim_{n\to\infty} s_n = R. \tag{3.1}$$

The rate of convergence measures how quickly the error $E_n = s_n - R$ reduces to zero, which is defined as

$$\lim_{n\to\infty} \frac{|E_{n+1}|}{|E_n|^q} = \lim_{n\to\infty} \frac{|s_{n+1} - R|}{|s_n - R|^q} = A, \quad A > 0, \quad q \geq 1, \tag{3.2}$$

where q represents the order of convergence of the iterative sequence [98, 121].

X.-S. Yang, X.-S. He, *Mathematical Foundations of Nature-Inspired Algorithms*, SpringerBriefs in Optimization, https://doi.org/10.1007/978-3-030-16936-7_3

In other words, we say that the sequence converges to R with the order of q. Here, A is called the asymptotic error constant or the rate of convergence.

- If $q = 1$ and $0 < A < 1$, we say the convergence is linear. The convergence is said to be superlinear if $A = 0$ and $q = 1$. In case of $A = 1$ and $q = 1$, the convergence is sublinear.
- If $q = 2$, the convergence is quadratic. We can also say that the sequence has a quadratic rate of convergence.
- If $q = 3$, the convergence is cubic.

For example, the following sequence

$$s_n = 2^{-n} \quad (n = 0, 1, 2, \ldots), \tag{3.3}$$

or

$$1, \frac{1}{2}, \frac{1}{4}, \frac{1}{8}, \frac{1}{16}, \ldots, \tag{3.4}$$

converges towards 0. That is, $\lim_{n\to\infty} s_n = R = 0$. Let us try

$$\lim_{n\to\infty} \frac{|s_{n+1} - R|}{|s_n - R|} = \lim_{n\to\infty} \frac{|2^{-(n+1)} - 0|}{|2^{-n} - 0|} = \lim_{n\to\infty} \frac{2^{-n}2^{-1}}{2^{-n}} = \frac{1}{2} = A. \tag{3.5}$$

Thus, $q = 1$ and $A = 1/2$, so the sequence converges linearly.

It is straightforward to show that 2^{-n^2} convergences to 0 superlinearly because

$$\lim_{n\to\infty} \frac{|2^{-(n+1)^2} - 0|}{|2^{-n^2} - 0|} = \lim_{n\to\infty} \frac{2^{-n^2}2^{-(2n+1)}}{2^{-n^2}} = \lim_{n\to\infty} 2^{-(2n+1)} = 0. \tag{3.6}$$

However, the sequence $s_n = 2^{-2^n}$ converges to 0 quadratically because

$$\lim_{n\to\infty} \frac{|2^{-2^{n+1}} - 0|}{|2^{-2^n} - 0|^2} = \lim_{n\to\infty} \frac{2^{-2^n\times 2}}{2^{-2^2\times 2}} = 1, \tag{3.7}$$

which gives $q = 2$ and $A = 1$.

3.1.2 Convergence Analysis of Newton's Method

Newton's method for optimization can be analysed from the point of view of root-finding algorithms. In essence, the iteration process is to find the root of the gradient $g(x) = f'(x) = 0$. Let us focus on the case of univariate functions, so the iterative algorithm can be written as

$$x_{k+1} = x_k - \frac{g(x_k)}{g'(x_k)}. \tag{3.8}$$

If we define

$$\phi(x) = x - \frac{g(x)}{g'(x)}, \tag{3.9}$$

we have

$$x_{k+1} = \phi(x_k). \tag{3.10}$$

Let R be the final solution as $k \to \infty$. Then we can expand ϕ around R and we have

$$\phi(x) = \phi(R) + \phi'(R)(x - R) + \frac{\phi''(R)}{2}(x - R)^2 + \cdots . \tag{3.11}$$

If we only consider the expansion up to the second order and using $\phi(R) = R$ (a fixed point), we have

$$x_{k+1} \approx R + \phi'(R)(x_k - R) + \frac{\phi''(R)}{2}(x_k - R)^2, \tag{3.12}$$

where we have used $x_{k+1} = \phi(x_k)$.

Since

$$\phi'(x) = \frac{g(x)g''(x)}{[g'(x)]^2}, \tag{3.13}$$

we know that there are two cases:

- If $\phi'(R) \neq 0$, we know that $x_k \to R$, so $(x_k - R)^2 \to 0$ or

  $$\frac{\phi''(R)}{2}(x_k - R)^2 \to 0, \quad k \to \infty,$$

  which means

  $$\lim_{k\to\infty} \frac{|x_{k+1} - R|}{|x_k - R|} = |g'(R)| \geq 0. \tag{3.14}$$

  The sequence thus converges linearly.

- If $\phi'(R) = 0$, the second-order term dominates and we have

$$\lim_{k\to\infty} \frac{|x_{k+1} - R|}{|x_k - R|^2} = \frac{|\phi''(R)|}{2}. \tag{3.15}$$

  This means that the convergence in this case is quadratic.

By the same philosophy, if $\phi'(R) = \phi''(R) = 0$, then the rate of convergence can be cubic if $\phi'''(R) \neq 0$.

For example, for $g(x) = x^2$, we have $g'(x) = 2x$ and $g''(x) = 2$, so

$$\phi'(x) = \frac{g(x)g''(x)}{[g'(x)]^2} = \frac{x^2 \times 2}{[2x]^2} = \frac{1}{2} \neq 0, \tag{3.16}$$

which means that the convergence is linear.

Interestingly, for $g(x) = x^2 - 1$, we know that its roots are ± 1. We know that $g'(x) = 2x$ and $g''(x) = 2$, thus

$$\phi'(x) = \frac{g(x)g''(x)}{[g'(x)]^2} = \frac{(x^2 - 1) \times 2}{[2x]^2} = 0, \tag{3.17}$$

where we have used $x^2 - 1 = 0$. Therefore, the iterative sequence converges to ± 1 quadratically.

3.2 Stability of an Algorithm

The stability of an algorithm is very important to ensure its accuracy; otherwise, the results may oscillate and the errors will increase, leading to wrong or meaningless solutions. Even for a relatively stable algorithm, the function values such as $\sin(x)/x$ may change significantly for very small $x \to 0$ such as $x \approx 10^{-10}$. Thus, care should be taken.

In general, an algorithm becomes unstable if the errors become large. It may be caused by rounding errors, cancellation errors such as $\sqrt{x^2 + a} - x$, where $|x| \gg |a|$, or the large modulus of the iteration formulas.

For example, if the error $E_k = |x_k - x_*|$ increases with iteration k in the case of

$$E_{k+1} = AE_k, \tag{3.18}$$

the system will diverge. Thus, some constraints (such as the condition number) on the form of the matrix $\boldsymbol{A}$ and its eigenvalues should be imposed so as to make the iterative algorithm stable [98, 118, 119].

In many applications, especially problems involved large matrices and the variations of quantities over many different magnitudes, some proper rescalings are needed so as to ensure stability.

3.3 Robustness Analysis

Robustness analysis was initially for robust engineering so as to produce robust designs under uncontrollable noise and uncertainties [51]. The aim is to produce best or better designs that are relatively insensitive to uncontrollable sources of variations or uncertainties such as materials properties in manufacturing and production. The performance is often evaluated in terms of some loss function or merit function, often as an expected value over different system configurations.

In the context of algorithm analysis, robustness can have two different aspects of meaning. Almost all algorithms have algorithm-dependent parameters, and the performance of an algorithm can largely depend on the setting of its parameters. Ideally, the algorithm should provide sufficiently good performance and such performance should be relatively insensitive to its parameter settings or small variations in parameter settings [101, 102]. This is one part of algorithmic robustness. On the other hand, for a given algorithm, the solutions (especially optimal or suboptimal solutions) obtained by an algorithm should also be relatively insensitive to the variations of initial configurations, the uncertainties in design parameters. Thus, the aim is to search for solutions that lie in the relatively smooth peak regions (for maximization) or valley regions (for minimization) of the objective landscape.

In some applications, both the design variables and the objective may contain noise or uncertainty. In this case, the expectation of the objectives is often used for optimization, while the constraints can also be modified in terms of their expectation when enough information (such as the probability distribution) is known.

For the statistical analysis of algorithms, robust statistics can be used, which often means to use non-parametric statistics, different norms and ranking methods. Though there are some good techniques for robust statistical analysis, however, the robust estimation in higher-dimensional spaces is extremely challenging if not impossible.

3.4 Probability Theory

Many algorithms nowadays use a certain degree of randomness, and thus probability theory will be useful to analyse such algorithms. We now briefly introduce random variables and probability distributions.

3.4.1 Random Variables

Randomness such as roulette-rolling and noise arises from the lack of information, or incomplete knowledge of reality. It can also come from the intrinsic complexity, diversity and perturbations of the system. Probability P is a number or an expected frequency assigned to an event A that indicates how likely it is that the event will occur when a random experiment is performed. This probability is often written as $P(A)$ to show that the probability P is associated with event A.

For a discrete random variable X with distinct values such as the number of cars passing through a junction, each value x_i may occur with a certain probability $p(x_i)$. In other words, the probability varies and is associated with its corresponding random variable. Traditionally, an uppercase letter such as X is used to denote a random variable, while a lowercase letter such as x_i is to represent its values. For example, if X means a coin-flipping event, then $x_i = 0$ (tail) or 1 (head). A probability function $p(x_i)$ is a function that assigns probabilities to all the discrete values x_i of the random variable X.

As an event must occur inside a sample space, the requirement is that all the probabilities must be summed to one, which leads to

$$\sum_{i=1}^{n} p(x_i) = 1. \tag{3.19}$$

For example, the outcomes of tossing a fair coin form a sample space. The outcome of a head (H) is an event with a probability of $P(H) = 1/2$, and the outcome of a tail (T) is also an event with a probability of $P(T) = 1/2$. The sum of both probabilities should be one, that is,

$$P(H) + P(T) = \frac{1}{2} + \frac{1}{2} = 1. \tag{3.20}$$

The cumulative probability function of X is defined by

$$P(X \le x) = \sum_{x_i \le x} p(x_i). \tag{3.21}$$

For a continuous random variable X that takes a continuous range of values (such as the level of noise), its distribution is continuous and the probability density function $p(x)$ is defined for a range of values $x \in [a, b]$ for given limits a and b [or even over the whole real axis $x \in (-\infty, \infty)$]. In this case, we always use the interval $(x, x + dx]$ so that $p(x)$ is the probability that the random variable X takes the value $x < X \le x + dx$ is

$$\Phi(x) = P(x < X \le x + dx) = p(x)dx. \tag{3.22}$$

As all the probabilities of the distribution shall be added to unity, we have

$$\int_a^b p(x)dx = 1. \tag{3.23}$$

The cumulative probability function becomes

$$\Phi(x) = P(X \leq x) = \int_a^x p(x)dx, \tag{3.24}$$

which is the definite integral of the probability density function between the lower limit a up to the present value $X = x$. Two main measures for a random variable X with a given probability distribution $p(x)$ are its mean and variance [119].

The mean μ or expectation of $\mathbb{E}[X]$ is defined by

$$\mu \equiv \mathbb{E}[X] \equiv <X> = \int xp(x)dx, \tag{3.25}$$

for a continuous distribution and the integration is within the integration limits. If the random variable is discrete, then the integration becomes the weighted sum

$$\mathbb{E}[X] = \sum_i x_i p(x_i). \tag{3.26}$$

The variance $\text{var}[X] = \sigma^2$ is the expectation value of the deviation squared $(X - \mu)^2$. That is,

$$\sigma^2 \equiv \text{var}[X] = \mathbb{E}[(X - \mu)^2] = \int (x - \mu)^2 p(x)dx. \tag{3.27}$$

The square root of the variance $\sigma = \sqrt{\text{var}[X]}$ is called the standard deviation, which is simply σ.

For a discrete distribution, the variance simply becomes the following sum:

$$\sigma^2 = \sum_i (x - \mu)^2 p(x_i). \tag{3.28}$$

In addition, any other formulas for a continuous distribution can be converted to their counterparts for a discrete distribution if the integration is replaced by the sum. Therefore, we will mainly focus on the continuous distribution in the rest of the section.

From the above definitions, it is straightforward to prove

$$\mathbb{E}[\alpha x + \beta] = \alpha\mathbb{E}[X] + \beta, \quad \mathbb{E}[X^2] = \mu^2 + \sigma^2, \tag{3.29}$$

and

$$\text{var}[\alpha x + \beta] = \alpha^2 \text{var}[X], \tag{3.30}$$

where α and β are constants.

Other frequently used measures are the mode and median. The mode of a distribution is defined by the value at which the probability density function $p(x)$ is the maximum. For an even number of data sets, the mode may have two values. The median m of a distribution corresponds to the value at which the cumulative probability function $\Phi(m) = 1/2$. The upper and lower quartiles Q_U and Q_L are defined by $\Phi(Q_U) = 3/4$ and $\Phi(Q_L) = 1/4$.

3.4.2 Poisson Distribution and Gaussian Distribution

The Poisson distribution is the distribution for small-probability discrete events. Typically, it is concerned with the number of events that occur in a certain time interval (e.g., number of telephone calls in an hour) or spatial area.

The probability density function of the Poisson distribution is

$$P(X = x) = \frac{\lambda^x e^{-\lambda}}{x!}, \quad \lambda > 0, \tag{3.31}$$

where $x = 0, 1, 2, \ldots, n$ and λ is the mean of the distribution.

Obviously, the sum of all the probabilities must be equal to one. That is,

$$\sum_{x=0}^{\infty} \frac{\lambda^x e^{-\lambda}}{x!} = \frac{\lambda^0 e^{-\lambda}}{0!} + \frac{\lambda^1 e^{-\lambda}}{1!} + \frac{\lambda^2 e^{-\lambda}}{2!} + \frac{\lambda^3 e^{-\lambda}}{3!} + \ldots$$

$$= e^{-\lambda}[1 + \lambda + \frac{\lambda^2}{2!} + \frac{\lambda^3}{3!} + \ldots] = e^{-\lambda} e^{\lambda} = e^{-\lambda+\lambda} = e^0 = 1. \tag{3.32}$$

Many stochastic processes such as the number of phone calls in a call centre and the number of cars passing through a junction obey the Poisson distribution. If we are concerned with a Poisson distribution with a time interval t, λ will be the arrival rate per unit time. However, in general, we should use $x = \lambda t$ to replace x when dealing with the arrivals in a fixed period t. Thus, the Poisson distribution becomes

$$P(X = n) = \frac{(\lambda t)^n e^{-\lambda t}}{n!}. \tag{3.33}$$

Using the definitions of mean and variance, it is straightforward to prove that $\mu = \lambda$ and $\sigma^2 = \lambda$ for the Poisson distribution.

The Gaussian distribution or normal distribution is the most important continuous distribution in probability and it has a wide range of applications. For a continuous random variable X, the probability density function (PDF) of the Gaussian distribution is given by

$$p(x) = \frac{1}{\sigma\sqrt{2\pi}} e^{-\frac{(x-\mu)^2}{2\sigma^2}}, \tag{3.34}$$

where $\sigma^2 = \text{var}[X]$ is the variance and $\mu = \mathbb{E}[X]$ is the mean of the Gaussian distribution. It is straightforward to verify that

$$\int_{-\infty}^{\infty} p(x)dx = 1, \tag{3.35}$$

and this is exactly the reason why the factor $1/\sqrt{2\pi}$ is required in the normalization of all the probabilities.

The probability function reaches a peak at $x = \mu$ and the variance σ^2 controls the width of the peak.

The cumulative probability function (CPF) for a normal distribution is the integral of $p(x)$, which is defined by

$$\Phi(x) = P(X < x) = \frac{1}{\sqrt{2\pi\sigma^2}} \int_{-\infty}^{x} e^{-\frac{(\zeta-\mu)^2}{2\sigma^2}} d\zeta, \tag{3.36}$$

which can be written as

$$\Phi(x) = \frac{1}{\sqrt{2}}\Big[1 + \text{erf}\Big(\frac{x-\mu}{\sqrt{2}\sigma}\Big)\Big], \tag{3.37}$$

where the error function is defined as

$$\text{erf}(x) = \frac{2}{\sqrt{\pi}} \int_0^x e^{-\zeta^2} d\zeta. \tag{3.38}$$

3.4.3 Common Probability Distributions

There are a number of other important distributions such as the exponential distribution, binomial distribution, Cauchy distribution, Lévy distribution and Student's t-distribution [35, 39, 47].

A Bernoulli distribution is a distribution of outcomes of a binary random variable X, where the random variable can only take two values: either 1 (success or yes) or 0 (failure or no). The probability of taking 1 is $0 \le p \le 1$, while the probability of taking 0 is $q = 1 - p$. Then, the probability mass function can be written as

$$B(m, p) = \begin{cases} p & \text{if } m = 1, \\ 1 - p, & \text{if } m = 0, \end{cases} \tag{3.39}$$

which can be written more compactly as

$$B(m, p) = p^m (1 - p)^{1-m}, \quad m \in \{0, 1\}. \tag{3.40}$$

It is straightforward to show that its mean and variance are

$$\mathbb{E}[X] = p, \quad \mathrm{var}[X] = pq = p(1-p). \tag{3.41}$$

This is the probability of a single experiment with two distinct outcomes. In case of multiple experiments or trials (n), the probability distribution of exactly m successes becomes the binomial distribution

$$B_n(m, n, p) = \binom{n}{m} p^m (1-p)^{n-m}, \tag{3.42}$$

where

$$\binom{n}{m} = \frac{n!}{m!(n-m)!} \tag{3.43}$$

is the binomial coefficient. Here, $n!$ is the factorial and $n = n(n-1)(n-2)\ldots 1$. For example, $5! = 5 \times 4 \times 3 \times 2 \times 1 = 120$. Conventionally, we set $0! = 1$.

It is also straightforward to verify that

$$\mathbb{E}[X] = np, \quad \mathrm{var}[X] = np(1-p), \tag{3.44}$$

for n trials.

The exponential distribution has the following probability density function:

$$f(x) = \lambda e^{-\lambda x}, \quad \lambda > 0, \quad (x > 0), \tag{3.45}$$

and $f(x) = 0$ for $x \leq 0$. Its mean and variance are

$$\mu = 1/\lambda, \quad \sigma^2 = 1/\lambda^2. \tag{3.46}$$

Cauchy probability distribution can be written as

$$p(x, \mu, \gamma) = \frac{1}{\pi\gamma}\Big[\frac{\gamma^2}{(x-\mu)^2 + \gamma^2}\Big], \quad -\infty < x < \infty, \tag{3.47}$$

its mean and variance are undefined or infinite, which is a true indication that this distribution is heavy-tailed. The cumulative distribution function of the Cauchy distribution is

$$F(x) = \frac{1}{\pi}\tan^{-1}\Big(\frac{x-\mu}{\gamma}\Big) + \frac{1}{2}. \tag{3.48}$$

It is worth pointing out that this distribution can have a heavy tail or a fat tail where probability can be still significantly non-zero at the tails as $x \to \infty$. Thus, such a distribution belongs to the heavy-tailed or fat-tailed distributions.

Other heavy-tailed distributions include Pareto distribution, power-law distributions and Lévy distribution.

A Pareto distribution is defined by

$$p(x) = \begin{cases} \frac{\alpha x_0^\alpha}{x^{\alpha+1}} & \text{if } x \geq x_0, \\ 1 & \text{if } x < x_0, \end{cases} \tag{3.49}$$

where $\alpha > 0$ and $x_0 > 0$ is the minimum value of x. The cumulative probability function is

$$F(x) = \begin{cases} 1 - \left(\frac{x_0}{x}\right)^\alpha & \text{for } x \geq x_0, \\ 0 & \text{for } x < x_0. \end{cases} \tag{3.50}$$

The power-law probability distribution can be written as

$$p(x) = Ax^{-\alpha}, \quad x \geq x_0 > 0, \tag{3.51}$$

where $\alpha > 1$ is an exponent and $A = (\alpha - 1)x_0^{\alpha-1}$ is the normalization constant. Alternatively, we can write the above as

$$p(x) = \frac{\alpha - 1}{x_0}\left(\frac{x}{x_0}\right)^{-\alpha}. \tag{3.52}$$

Lévy probability distribution is given by

$$p(x, \mu, \gamma) = \frac{\sqrt{\frac{\gamma}{2\pi}}}{(x - \mu)^{3/2}} e^{-\frac{\gamma}{2(x-\mu)}}, \quad x \geq \mu, \tag{3.53}$$

where $\mu > 0$ controls its location and γ controls its scale [88]. For a more general case with an exponent β, we have to use the integral to define Lévy distribution

$$p(x) = \frac{1}{\pi} \int_0^\infty \cos(kx) e^{-\alpha |k|^\beta} dk, \quad (0 < \beta \leq 2), \tag{3.54}$$

where $\alpha > 0$. The special case of $\beta = 1$ becomes a Cauchy distribution, while $\beta = 2$ corresponds to a normal distribution.

The Student's t-distribution is given by

$$p(t) = K\left(1 + \frac{t^2}{n}\right)^{-(n+1)/2}, \quad -\infty < t < +\infty, \tag{3.55}$$

where n is the number of degrees of freedom, and

$$K = \frac{\Gamma((n+1)/2)}{\sqrt{n\pi}\, \Gamma(n/2)}, \tag{3.56}$$

where Γ is the gamma function defined by

$$\Gamma(z) = \int_0^\infty x^{z-1} e^{-x} dx. \tag{3.57}$$

When $z = n$ is an integer, we have $\Gamma(n) = (n-1)!$.

It is worth pointing out that Pareto distribution, power-law distribution and Lévy distributions are one-tailed because they are valid for $x > x_{\min}$. On the other hand, both the Cauchy distribution and Student's t-distribution are two-tailed as they are valid for the whole real domain.

3.5 Random Walks and Lévy Flights

If a step size S_i is drawn from a normal distribution with zero mean and unit variance, we have

$$S_i \sim \frac{1}{\sqrt{2\pi}} \exp\Big[-\frac{x^2}{2}\Big], \tag{3.58}$$

where we have used '$\sim$' to mean 'draw random numbers that obey the distribution'. Then, the sum of N consecutive steps forms a random walk

$$W_N = \sum_{i=1}^{N} = S_1 + S_2 + \ldots + S_N, \tag{3.59}$$

which is equivalent to

$$W_N = W_{N-1} + S_N. \tag{3.60}$$

This means that the next step W_N (or state) depends only on the existing state or step W_{N-1} and the transition move (step) S_N. This kind of random walks can also be considered as a diffusion process or Brownian motion.

Loosely speaking, Lévy flights are a random walk whose step sizes are drawn from a Lévy distribution [88, 118], often in terms of a simple power-law formula $L(s) \sim |s|^{-1-\beta}$, where $0 < \beta \leq 2$ is an index. Mathematically speaking, a simple version of Lévy distribution can be defined as

$$L(s, \gamma, \mu) = \begin{cases} \sqrt{\frac{\gamma}{2\pi}} \exp\Big[-\frac{\gamma}{2(s-\mu)}\Big] \frac{1}{(s-\mu)^{3/2}}, & 0 < \mu < s < \infty, \\ 0, & \text{otherwise,} \end{cases} \tag{3.61}$$

where $\mu > 0$ is a minimum step and γ is a scale parameter. Clearly, as $s \to \infty$, we have

$$L(s, \gamma, \mu) \approx \sqrt{\frac{\gamma}{2\pi}} \frac{1}{s^{3/2}}. \tag{3.62}$$

This is a special case of the generalized Lévy distribution.

In general, Lévy distribution should be defined in terms of the following Fourier transform:

$$F(k) = \exp[-\alpha |k|^{\beta}], \quad 0 < \beta \leq 2, \tag{3.63}$$

where α is a scale parameter. The inverse of this integral is not easy, as it does not have any analytical form, except for a few special cases.

For the case of $\beta = 2$, we have

$$F(k) = \exp[-\alpha k^2], \tag{3.64}$$

whose inverse Fourier transform corresponds to a Gaussian distribution. Another special case is $\beta = 1$, and we have

$$F(k) = \exp[-\alpha |k|], \tag{3.65}$$

which corresponds to a Cauchy distribution

$$p(x, \gamma, \mu) = \frac{1}{\pi} \frac{\gamma}{\gamma^2 + (x - \mu)^2}, \tag{3.66}$$

where μ is the location parameter, while γ controls the scale of this distribution.

For the general case, the inverse integral

$$L(s) = \frac{1}{\pi} \int_0^{\infty} \cos(ks) \exp[-\alpha |k|^{\beta}] dk \tag{3.67}$$

can be estimated only when s is large. We have

$$L(s) \to \frac{\alpha \, \beta \, \Gamma(\beta) \sin(\pi \beta / 2)}{\pi |s|^{1+\beta}}, \quad s \to \infty. \tag{3.68}$$

Here $\Gamma(z)$ is the Gamma function defined earlier.

Lévy flights are more efficient than Brownian random walks in exploring unknown, large-scale search space. There are many reasons to explain this efficiency, and one of them is due to the fact that the variance of Lévy flights

$$\sigma^2(t) \sim t^{3-\beta}, \quad 1 \leq \beta \leq 2, \tag{3.69}$$

increases much faster than the linear relationship (i.e., $\sigma^2(t) \sim t$) of Brownian random walks [79].

It is worth pointing out that a power-law distribution is often linked to some scale-free characteristics, and Lévy flights can thus show self-similarity and fractal behaviour in the flight patterns.

From the implementation point of view, the generation of random numbers with Lévy flights consists of two steps: the choice of a random direction and the generation of steps which obey the chosen Lévy distribution. The generation of a direction should be drawn from a uniform distribution, while the generation of steps is quite tricky. There are a few ways of achieving this, but one of the most efficient and yet straightforward ways is to use the so-called Mantegna's algorithm for a symmetric Lévy stable distribution. Here 'symmetric' means that the steps can be positive and negative.

In Mantegna's algorithm [65], the step length s can be calculated by

$$s = \frac{u}{|v|^{1/\beta}}, \tag{3.70}$$

where u and v are drawn from normal distributions. That is,

$$u \sim N(0, \sigma_u^2), \quad v \sim N(0, \sigma_v^2), \tag{3.71}$$

where

$$\sigma_u = \left\{ \frac{\Gamma(1+\beta) \sin(\pi\beta/2)}{\Gamma[(1+\beta)/2] \, \beta \, 2^{(\beta-1)/2}} \right\}^{1/\beta}, \quad \sigma_v = 1. \tag{3.72}$$

This distribution (for s) obeys the expected Lévy distribution for $|s| \geq |s_0|$, where s_0 is the smallest step. In principle, $|s_0| \gg 0$, but in reality s_0 can be taken as a sensible value such as $s_0 = 0.1$–1 [118].

3.6 Performance Measures

There are many different performance measures used in the literature, and they can be summarized as the following five categories:

- *Accuracy*: A very commonly used performance measure is to see the accuracy of the solutions that an algorithm can obtain. For example, if one uses an algorithm to find the roots of $f(x) = x^2 - 4 = 0$ for a fixed number of ten iterations starting with an initial value $x = 10$, Newton's method can give a higher accuracy than the bisection method. In this case, Newton's method can be a better choice.

  The accuracy is often expressed as the decimal places or the tolerance, depending on the context. For example, for a minimization problem $f(x) = x^2$, an algorithm (Algorithm A) may get $f_{\min} = 1.23 \times 10^{-16}$, while the other algorithm (Algorithm B) may get $f_{\min} = 2.40 \times 10^{-23}$. In this case, we can say that B obtains a higher accuracy.

However, to compare accuracy fairly, we have to ensure computational efforts such as the number of function evaluations must be same. Otherwise, if you run one algorithm much longer than other algorithm, ultimately it may produce a higher accuracy even though it may not be a good algorithm.

- *Function Evaluations*: Computational efforts can be measured as the number of function evaluations. In most cases, the evaluations of the objective functions can be expensive, which is especially true in engineering and industrial applications as well as bioinformatics where the evaluation of a single call of the objective function can take hours or even days.

  In order to obtain the same accuracy (say to the fifth decimal place or 10^{-5}), if algorithm A takes 100 iterations, while algorithm B take 2000 iterations, we can say A is much more efficient than B.
- *Complexity*: The time complexity of an algorithm can be analysed. For most nature-inspired algorithms with a population size of n and the number of iterations t, the computational complexity is often the order of $O(nt)$ or $O(n^2 t)$, depending on the ways of looping over the whole population. The spatial complexity is typically $O(n)$ or at most $O(n^2)$ for storage.

  However, the actual computation time on a computer is not a good indicator because the actual execution time will depend on many factors such as the hardware configuration, the software used, the exact implementation, operating system used and other software packages running or hidden in the background. For example, computational times will be affected by most anti-virus software packages. In addition, the vectorization such as in Matlab and C++ will run significantly faster than a `for` loop.
- *Success Rate*: Many researchers use the rate of successful runs as a performance indicator. For example, if the true global minimum of a minimization problem is $f_{\min} = 0$, and if an algorithm finds f such that $|f - f_{\min}| \le \delta$, where $\delta = 10^{-5}$ or 10^{-10}, we can consider it as a success. Typically, if there are N independent runs, if $M \le N$ runs find a solution that is sufficiently close to the true solution, the success rate can be defined as

$$S_r = \frac{M}{N}, \tag{3.73}$$

  which is often expressed as a percentage. However, the success rate will depend on the value of tolerance δ. For a success run with $\delta = 10^{-5}$, it may not be a success for $\delta = 10^{-10}$ or even $\delta = 10^{-6}$. Therefore, to compare two algorithms, the success rate must be calculated in the same ways so as to compare them fairly.

  In addition, the success rate can also be defined in terms of the closeness of the solution in the search space. If the true optimal solution is $\boldsymbol{x}_*$, a solution $\boldsymbol{x}$ found by an algorithm is called a success if $|\boldsymbol{x} - \boldsymbol{x}_*| \le \delta$. This definition is obviously different from $|f(x) - f_{\min}(\boldsymbol{x}*)| \le \delta$. In case of multiple multimodal problems, a single value of $|f(x) - f_{\min}|$ may correspond to multiple values of $|\boldsymbol{x} - \boldsymbol{x}_*|$, and thus care should be taken when dealing with the success rates.

- *Diversity and Robustness*: In some applications, diversity of the solutions can be very important, which is true for multiobjective optimization where the diversity of solutions may enable a more uniform distribution of solution points on the Pareto front. One way of measuring the diversity of the population or solutions is to use the variance or standard deviation, in addition to the mean, the best and worse solutions. For the solutions on a Pareto front, their diversity is measured by crowd distance or crowding distance to show how many solutions or points in a neighbourhood of an often fixed size.

  Robustness is more subtle to measure. Some researchers use sensitivity to see how a small variation of a parameter may affect the variations of solutions, while others use stability of the system to small perturbations. Some researchers even use slightly different meaning to mean robustness, depend on the context of usage. For example, the uncertainty in materials properties and their effect on the final design solutions may also be considered as robustness. In fact, there is a subject on robustness analysis with extensive literature [69].

It is worth pointing out that due to the stochastic nature of many algorithms used in optimization, data mining and machine learning, multiple runs are needed to estimate the performance measures. In fact, such performance measures may behave like a random variable, and thus their mean, standard deviation, the best and worse values should be given so as to give a fuller picture of the evaluations of the algorithms under consideration.

In addition, some statistical testing such as Student's t-test, Wilcoxon signed-rank test and others can be used to see if there is any significant difference in terms of different performance measures between any two algorithms.

3.7 Monte Carlo and Markov Chains

In many applications, the number of possible combinations and states is so astronomical that it is impossible to carry out evaluations over all possible combinations systematically. In this case, the Monte Carlo method is one of the best alternatives. Monte Carlo is in fact a class of methods now widely used in computer simulations, machine learning and weather forecasting. Since the pioneer studies in the 1940s and 1950s, especially the work by Ulam, von Newmann and Metropolis, it has been applied in almost all areas of simulations, from numerical integration and Internet routing to financial market and climate simulations [35, 43].

The error of Monte Carlo integration decreases with N in a manner of $1/\sqrt{N}$, which is independent of the number (D) of dimensions. This becomes advantageous over other conventional methods for multiple integrals in higher dimensions.

The basic procedure is to generate the random points so that they distribute uniformly inside the domain. In order to calculate the volume in higher dimensions, it is better to use a regular control domain to enclose the domain Ω of the integration.

Markov chains are a more general framework where the states of a system evolve, but the next states only depend on the current states and their transition probabilities. The simple random walks we discussed earlier in this chapter are a Markov chain. Similarly, Lévy flights are also Markov chains. An important property of Markov chains is that the long-term behaviour will not depend on the starting points. In other words, the states in the long run will 'forget' their initial states, and thus settle onto some stable distributions [35, 39].

The advance of Markov chain Monte Carlo (MCMC) provides a possible framework for stochastic optimization methods [43]. In essence, MCMC is a class of sample-generating methods, and it attempts to draw samples directly from some highly complex distributions using Markov chains with known transition probabilities. This method has become a powerful tool for Monte Carlo simulations, Bayesian statistical analysis and optimization.

Loosely speaking, almost all metaheuristic algorithms (including nature-inspired algorithms) can be considered as some form of Monte Carlo approaches, though selection mechanisms used in such algorithms make the Monte Carlo sampling biased towards some potentially promising regions in the search space. For example, simulated annealing generates solution points randomly, and the selection in terms of a probability makes its search path zig-zag towards potentially optimal regions.

Chapter 4
Mathematical Analysis of Algorithms: Part I

With the introduction of some of the major nature-inspired algorithms and the brief outline of mathematical foundations, now we are ready to analyse these algorithms in great detail. Obviously, we can analyse these algorithms from different angles and perspectives. Let us first analyse the basic components and mechanisms of most algorithms so as to gain some insights.

4.1 Algorithm Analysis and Insight

Though we know that all the above algorithms we have discussed and other algorithms can work well in practice and they are able to solve a diverse range of problems, we rarely understand how they work exactly. To gain a truly comprehensive, in-depth understanding of all algorithms, it requires a multidisciplinary approach by combining mathematical analysis, numerical analysis, computational complexity, dynamical systems and other relevant tools. Therefore, we will not attempt such challenging tasks here in this section. Instead, we will first focus on analysing the basic characteristics of algorithms, their components, search mechanisms and behaviour so as to gain a better insight into such algorithms. After such qualitative analysis in this section, we will try to provide some mathematical analyses in the next section.

4.1.1 Characteristics of Nature-Inspired Algorithms

First, let us look at nature-inspired algorithms by their basic steps, search characteristics and algorithm dynamics.

X.-S. Yang, X.-S. He, *Mathematical Foundations of Nature-Inspired Algorithms*, SpringerBriefs in Optimization, https://doi.org/10.1007/978-3-030-16936-7_4

- All algorithms use a population of multiple agents (e.g., particles, ants, bats, cuckoos, fireflies, bees, etc.), each agent corresponds to a solution vector. Among the population, there is often the best solution $\boldsymbol{g}^*$ in terms of objective fitness. Different solutions in a population represent both diversity and different fitness.
- The evolution of the population is often achieved by some operators (e.g., mutation and crossover), often in terms of some algorithmic formulas or equations. Such evolution is typically iterative, leading to evolution of solutions with different properties. When all solutions become sufficiently similar, the system can be considered as converged.
- The moves of an agent represent a zigzag piecewise path in the search space, and such moves are quasi-deterministic. Thus, randomization techniques are often used to generate new solution vectors or moves. Such randomization provides a mechanism to perturb the states (or solutions) of the algorithm, which potentially allows it to escape any local optima (thus minimizing the probability of getting stuck locally).
- All algorithms try to carry out some sort of both local and global search. If the search is mainly local, it increases the probability of getting stuck locally. If the search focuses too much on global moves, it will slow down the convergence. Different algorithms may use different amount of randomization and different portion of moves for local or global search.

  However, it is not clear yet what the right amount of randomness is and what the ratio of global search to local search should be.
- Selection of the better or best solutions is carried out by the 'survival of the fittest' or simple elitism so that the best solution $\boldsymbol{g}^*$ is kept in the population in the next generation. Such selection essentially acts as a driving force to drive the diverse population into a converged population with reduced diversity but with a more organized structure.

These basic components, characteristics and their properties can be summarized in Table 4.1, and this can form a basis for comparison with the characteristics of self-organization to be discussed later.

Table 4.1 Characteristics of nature-inspired algorithms

Components/characteristics	Role or properties
Population	Diversity and sampling
Randomization/perturbations	Escape local optima
Selection and elitism	Driving force for convergence
Algorithmic equations	Iterative evolution of solutions

4.1.2 What's Wrong with Traditional Algorithms?

One may wonder what is wrong with traditional algorithms? A short answer is that there is nothing wrong. Traditional algorithms work well for the types of problems they can solve, but most traditional algorithms are *local search*.

- As traditional algorithms are mostly local search, there is no guarantee for global optimality for most optimization problems, except for linear programming and convex optimization. Consequently, the final solution will often depend on the initial starting points (except for linear programming and convex optimization).
- Traditional algorithms tend to be problem-specific because they usually use some information such as derivatives about the local objective landscape. Other methods such as k-opt, branch and bound and others can heavily depend on the type of problems in implementation.
- Traditional algorithms cannot solve highly nonlinear, multimodal problems effectively, and they often struggle to cope with problems with discontinuity, especially when gradients needed are expensive to calculate or estimate.
- Almost all traditional algorithms, except for hill-climbing with random restart, are deterministic algorithms. The final solutions will be identical if starting with the same initial points. No random numbers are used. Consequently, the diversity of the obtained solutions can be very limited.

4.2 Advantages of Heuristics and Metaheuristics

In order to remedy the above disadvantages, contemporary algorithms tend to be heuristic and metaheuristic. Heuristic algorithms use a trial-and-error approach in generating new solutions, while metaheuristic algorithms are a higher-level heuristic with the additional use of memory, solution history and other forms of 'learning' strategy. Nowadays, most metaheuristic algorithms are nature-inspired algorithms and most such algorithms are based on swarm intelligence inspired by nature [34, 99, 118]. In contrast with traditional algorithms, metaheuristics are mainly designed for *global search* and tend to have the following advantages and characteristics:

- As they are global optimizers, it is more likely to find the true global optimality.
- They often treat problems as a black box without specific knowledge, thus they can solve a wider range of problems.
- Metaheuristic algorithms are usually gradient-free methods and they do not use any derivative information, and thus can deal with highly nonlinear problems and problems with discontinuity.
- Stochastic components in terms of random numbers and random walks are often used, and thus such algorithms are stochastic. Thus, no identical solutions can be obtained, even starting with the same initial points, but the final solutions can be sufficiently close and they often enable the algorithm to escape any local modes (thus less likely to get stuck in local regions).

Despite these advantages, nature-inspired algorithms do have some disadvantages. In general, the computational efforts are higher than those for traditional algorithms because more iterations are needed, which can become too computationally expensive if the evaluation of a single objective requires a long time by a simulator (e.g., by finite element methods). In addition, the final solutions obtained by such algorithms cannot be repeated exactly, and multiple runs should be carried out to ensure consistency and some meaningful statistical analysis.

4.3 Key Components of Algorithms

Now let us look at the key components of nature-inspired algorithms more closely so as to understand their role and effects on the performance of algorithms.

4.3.1 Deterministic or Stochastic

A key feature of traditional algorithms is that they are mainly deterministic and no randomness is used in generating new solutions. This can enhance the exploitation ability, but lacks exploration capabilities. On the other hand, nature-inspired metaheuristic algorithms use a certain degree of randomness, and these algorithms have stochastic components. A good degree of randomness will increase the exploration ability, but may reduce the exploitation abilities.

Some questions arise naturally: Which is better? How much randomness should an algorithm have? As we discussed earlier, both traditional deterministic algorithms and stochastic metaheuristic algorithms have some advantages and disadvantages. From the global optimization perspective, the advantages of stochastic algorithms far outweigh their disadvantages. Both empirical observations and simulation suggest that randomness can be largely beneficial to the overall performance of algorithms. As to the right degree of randomness, it is very difficult to say because such randomness can depend on the algorithmic structure, type of problems and the solution quality desired for a given type of problems. In fact, this is still an open problem.

4.3.2 Exploration and Exploitation

Another way of analysing algorithms is to look at their exploration and exploitation abilities. Exploration provides diversification, which allows the algorithm to search different regions in the design space and thus increases the probability of finding the true global optimality.

Exploration is often achieved by randomization or random numbers in terms of some predefined probability distributions [15]. In most cases, random numbers drawn from a uniform distribution or a Gaussian distribution are used. Exploration can be considered as a global, explorative mechanism. For example, cuckoo search has a strong ability of exploration due to the use of Lévy flights. PSO uses two uniformly distributed random numbers to enable its exploration.

On the other hand, exploitation uses local information such as gradients to search local regions more intensively, and such intensification can enhance the rate of convergence. Exploitation can make the population less diverse, and strong local guidance can even make the population relatively uniform in terms of solution variations. For example, in PSO and bat algorithm, the best solution g^* is used to exploit the current best solution and its locality in the design space.

Too much exploration and too little exploitation can slow down the convergence of an algorithm, while too much exploitation and too little exploration can sacrifice the possibility of finding true global solutions. Therefore, there is a fine balance between exploration and exploitation, which may depend on the algorithmic structure and type of problems.

The estimated exploration and exploitation capabilities of some commonly used algorithms are shown in Figure 4.1. For example, gradient-based methods have high exploitation capabilities with very limited exploration, while uniform search has a very high exploration capability with almost no exploitation.

4.3.3 Role of Components

Alternatively, we can also analyse the algorithm components in terms of their role. Borrowing the terminologies from genetic algorithms, we can look at mutation, crossover and selection.

Most algorithms use mutation. For example, differential evolution uses a vectorized mutation operator $(\boldsymbol{x}_j - \boldsymbol{x}_k)$, and firefly algorithm uses an isotropic random walk. Many other algorithms such as PSO and bat algorithm use vectorized mutation

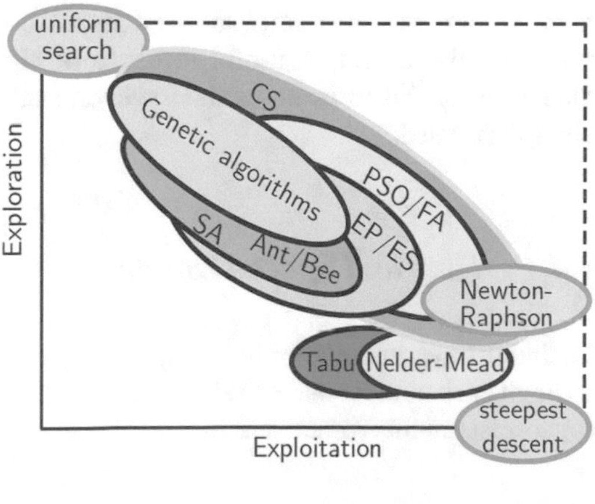

Fig. 4.1 Exploration and exploitation of some commonly used algorithms

in a similar way as that in differential evolution. However, cuckoo search and flower pollination algorithm use Lévy flights in terms of non-isotropic random walks, which makes the algorithms more efficient due to the power-law, scale-free search properties of Lévy flights.

Crossover is a mechanism that can enhance the mixing ability of the population, but not all algorithms use crossover. For example, differential evolution uses binomial and exponential crossover, but PSO, bat algorithm, cuckoo search and others do not use crossover explicitly. However, many variants of PSO, cuckoo search and flower pollination algorithm introduced some form of crossover, and they achieved enhanced performance.

Selection is a driving mechanism to ensure convergence among the populations. All algorithms have to have some good selection mechanisms. Genetic algorithms use elitism and survival of the fittest, while PSO uses both the best solution $\boldsymbol{g}^*$ and individual best $\boldsymbol{x}_i^*$ as selection. Firefly algorithm uses the brightest fireflies implicitly as an attraction mechanism. Other algorithms such as cuckoo search do not use $\boldsymbol{g}^*$, while flower pollination algorithm uses $\boldsymbol{g}^*$ explicitly. The use of $\boldsymbol{g}^*$ is something like a double-edged sword. If the selection mechanism is too strong, the diversity of the population can be limited. For example, in PSO, the use of both $\boldsymbol{g}^*$ and individual best solutions may be too strong for some problems, and the solutions can get stuck at some local regions, leading to potential premature convergence in this case. On the other hand, if the selection is weak, many solutions are not well-selected, and the convergence may be significantly slowed down. Again it needs a fine balance of selection strength as well as a good combination of crossover and mutation.

4.4 Complexity

To measure how easy or hard that a problem can be solved, we need to estimate its computational complexity [6]. We cannot simply ask how long it takes to solve a particular problem instance because the actual computational time will depend on both the hardware and software used to solve it. Thus, time does not make much sense in this context. A useful measure of complexity should be independent of the hardware and software used. However, such complexity is closely linked to the algorithms used.

4.4.1 Time and Space Complexity

To find the maximum (or minimum) among n different numbers, we only need to go through each number once by simply comparing the current number with the highest (or lowest) number once and update the new highest (or lowest) when necessary.

Thus, the number of mathematical operations is simply $O(n)$, which is the time complexity of this problem.

In practice, comparing two big numbers may take slightly longer, and different representations of numbers may also affect the speed of this comparison. In addition, multiplication and division usually take more time than simple addition and subtraction. However, in computational complexity, we usually ignore such minor differences and simply treat all operations as equal. In this sense, the complexity is about the number or order of mathematical operations, not the actual order of computational time.

On the other hand, space computational complexity estimates the size of computer memory needed to solve the problem. In the above simple problem of finding the maximum or minimum among n different numbers, the memory needed is $O(n)$ because it needs at least n different entries in the computer memory to store n different numbers. Though we need one more entry to store the largest or smallest number, this minor change does not affect the order of complexity because we implicitly assume that n is sufficiently large [6].

In most literature, if there is no time or space explicitly used when talking about computational complexity, it usually means time complexity. In discussing computational complexity, we often use the word 'problem' to mean a class of problems of the same type, and an 'instance' to mean a specific example of a problem class. Thus, $\boldsymbol{A}x = \boldsymbol{b}$ is a problem (class) for linear algebra, while

$$\begin{pmatrix} 2 & 3 \\ 1 & 1 \end{pmatrix} \begin{pmatrix} x \\ y \end{pmatrix} = \begin{pmatrix} 8 \\ 3 \end{pmatrix} \tag{4.1}$$

is an instance. In addition, a decision problem is a yes–no problem where an output is binary (0 or 1), even though the inputs can be any values.

Complexity classes are often associated with the computation on a Turing machine. Informally speaking, a Turing machine is an abstract machine that reads an input (a bit or a symbol) and manipulates one symbol at a time, following a set of fixed rules, by shifting a tape to the left or right. It can also write a symbol or bit to the tape (or scratch pad) which has an unlimited memory capacity. This basic concept can be extended to a universal Turing machine where the input and output can be represented as strings (either binary strings or strings in general), the tape or scratch pad can be considered as memory, and the manipulation as a transition function such as a simple Boolean function. It has been proved that a Turing machine is capable of carrying out the computation of any computable function. In this case, a universal Turing machine can be considered as a general-purpose modern computer with the transition function being the central processing unit, the tape being the computer memory and the inputs/outputs being the string representations. As the rules are fixed and at most one action is allowed at any given time, such a Turing machine is called a deterministic Turing machine.

In a more general case, a Turing machine has a set of rules that may allow more than one action, and the manipulation of symbols can have multiple branches, depending on current states. Thus, the transitions on the machine are not deterministic and multiple decision structures exist in this case. Such a Turing machine is called a non-deterministic Turing machine.

4.4.2 Complexity of Algorithms

The above two types of computational complexity are closely linked to the type of problems. Even for the same type of problem, different algorithms can be used, and the number of basic mathematical operations may be different. In this case, we are concerned about the complexity of an algorithm in terms of arithmetic complexity.

Computational complexity can be estimated for each algorithm, and this can help to under the computational efforts needed. For example, most nature-inspired algorithms such as PSO, FPA and bat algorithm have a complexity of $O(nT)$, where n is the population size and T is the total number of iterations. Firefly algorithm has a computational complexity of $O(n^2T)$. Since n is relatively small compared with T, this usually does not increase the computation efforts substantially. In general, the computational complexity of nature-inspired algorithms is low.

On the other hand, the complexity of problems to be solved can be very high, even non-deterministic polynomial-time (NP) hard. For example, the well-known travelling salesman problems are NP-hard. Even nature-inspired algorithms are relatively simple, studies have indicated that they can solve complex problems and even NP-hard problems. It still remains a bit mystery how such algorithms with low algorithmic complexity can solve highly complex problems and be able to find good solutions and even optimal solutions in practice.

4.5 Fixed Point Theory

Numerical analysis often places emphasis on the iterative nature of an algorithm $A(\boldsymbol{x}_t)$ and tries to figure out how the solution $\boldsymbol{x}_t$ sequence may evolve as a pseudo-time iteration counter t gradually increases. From the discussion in Chapter 2, we know that an algorithm can be written as

$$\boldsymbol{x}^{t+1} = A(\boldsymbol{x}^t, \boldsymbol{x}_*, p_1, \ldots, p_K), \tag{4.2}$$

for $t \geq 0$. The new solution $\boldsymbol{x}^{t+1}$ in an algorithm A will largely depend on the current solution $\boldsymbol{x}^t$, its historical best $\boldsymbol{x}_*$ and some algorithm-dependent parameters $(p_1, \ldots, p_K)$.

As the iteration continues, it is possible that

$$\lim_{t \to \infty} \boldsymbol{x}^{t+1} = \boldsymbol{x}_\infty, \tag{4.3}$$

where $\boldsymbol{x}_\infty$ is a fixed point. Obviously, if $\boldsymbol{x}_\infty$ does not exist, we can say the algorithm diverges. In a special case when $\boldsymbol{x}_\infty = \boldsymbol{x}_*$, we can safely say that the algorithm has found the true optimal solution, though multiple optimal solutions may exist for multimodal problems. But if $\boldsymbol{x}_\infty \neq \boldsymbol{x}_*$, it may indicate that the iteration sequence becomes prematurely converged.

It is worth pointing out that the above solutions usually have some randomness and noise for metaheuristic algorithms, and, therefore, the above equation should be interpreted as the mean or expectation. That is,

$$\mathbb{E}[\lim_{t\to\infty} \boldsymbol{x}^{t+1}] = < \lim_{t\to\infty} \boldsymbol{x}^{t+1} > = < \boldsymbol{x}_\infty > . \tag{4.4}$$

The general fixed point theory dictates how an iterative formula may evolve and lead to a fixed point in the search space [98]. It is worth pointing out that there may be multiple fixed points, and each iteration sequence may only find one fixed point at a time, though it is possible for some algorithms such as the firefly algorithm to find multiple fixed points simultaneously.

For a population of solutions in any nature-inspired algorithms, the population can interact with each other and may lead to potentially multiple fixed points, depending on the algorithm dynamics of each algorithm. It can be expected that the ultimate best solution $\boldsymbol{g}^*$ (not the best at each iteration) acts as a fixed point in PSO, while there may be multiple fixed points in the firefly algorithm. Therefore, we can hypothesize that there is a single fixed point in BA, PSO, simulated annealing, FPA and bee algorithm, while multiple fixed points can exist in FA, CS, ACO and genetic algorithms if the conditions are right. However, it is not clear yet what these conditions can be and how to maintain these conditions in practice. In addition, these conditions may also be problem-dependent. It is highly necessary to carry out more research in this area.

4.6 Dynamical System

The first analysis of PSO using a dynamical system theory was carried out by Clerc and Kennedy [25], and they linked the governing equations of PSO with the dynamical behaviour of particles under different parameter settings. Using matrix algebra, we can rewrite Equations (2.18) and (2.19) as the following dynamic system:

$$\begin{pmatrix} \boldsymbol{x}_i \\ \boldsymbol{v}_i \end{pmatrix}^{t+1} = \begin{pmatrix} 1 & 1 \\ -(\alpha\epsilon_1 + \beta\epsilon_2) & 1 \end{pmatrix} \begin{pmatrix} \boldsymbol{x}_i \\ \boldsymbol{v}_i \end{pmatrix}^{t} + \begin{pmatrix} 0 \\ \alpha\epsilon_1 \boldsymbol{g}^* + \beta\epsilon_2 \boldsymbol{x}_i^* \end{pmatrix}. \tag{4.5}$$

In their original analysis [25], they did not use the eigenvalues of the above system because the matrices contain random numbers. Instead, they made additional assumptions and their analysis suggested that the PSO system is governed by the eigenvalues of a system matrix

$$\lambda_{1,2} = 1 - \frac{\gamma}{2} \pm \frac{\sqrt{\gamma^2 - 4\gamma}}{2}, \tag{4.6}$$

which leads to a bifurcation at $\gamma = \alpha + \beta = 4$. This kind of analysis can indeed provide some insight into the working mechanism and main characteristics, but it may be difficult to provide a full picture of the system because of simplifications used in the analysis.

For the bat algorithm, we can rewrite the algorithmic equations as

$$\begin{pmatrix} \boldsymbol{x}_i \\ \boldsymbol{v}_i \end{pmatrix}^{t+1} = \begin{pmatrix} 1 & 1 \\ f_i & 1 \end{pmatrix} \begin{pmatrix} \boldsymbol{x}_i \\ \boldsymbol{v}_i \end{pmatrix}^{t} + \begin{pmatrix} 0 \\ f_i \end{pmatrix}, \tag{4.7}$$

where $f_i = f_{\min} + (f_{\max} - f_{\min})\beta$. A quick look seems to show that this system is very similar to (4.5), but we have not considered the variations of the pulse emission rate r and loudness A in the above equations. The similarity allows to do some similar analysis, but the incompleteness of this system to capture the full functionalities of the bat algorithm means that the analysis may not provide much useful information in practice.

For detailed discussions about the analysis of the bat algorithm using such dynamical system theory, readers can refer to a recent paper by Chen et al. [24] where a triangular region of parameter settings in the parameter space has been obtained.

In principle, we can use the similar method to analyse other algorithms; however, it becomes difficult to extend to a generalized system. For example, in FA, CS and ACO, the nonlinearity makes it difficult to figure out the eigenvalues because the matrix will depend on the current solution, randomization and other factors. Furthermore, nonlinearity in algorithms such as FA also means that the characteristics can be much richer than simple linear dynamics such as PSO. Thus, this method may become intractable in practice, and some linearization and approximations may be needed.

4.7 Self-organized Systems

Another way of looking at nature-inspired algorithms is from the perspective of self-organization. Loosely speaking, a complex system can self-organize when the size of the system is sufficiently large with a high number of degrees of freedom, perturbations and a driving mechanism, giving enough time for the system to evolve from noise and far from equilibrium states [8, 56]. Mathematically speaking, a

Table 4.2 Self-organization and algorithms

Self-organization	Characteristics	Algorithm	Properties
States	Complexity	Population	Diversity and sampling
Noise, perturbations	Far from equilibrium	Randomization	Escape local optima
Selection mechanism	Organization	Selection	Convergence
Re-organization	State changes	Iteration	Evolution of solutions

system with multiple states S_i can evolve with time t towards the self-organized states S_*, driven by a driving mechanism M which can be written schematically as

$$S_i \stackrel{M}{\Longrightarrow} S_*. \tag{4.8}$$

Now let us look at an algorithm using self-organization, an algorithm can indeed be considered as a self-organization system, starting from a population of solutions $\boldsymbol{x}_i (i = 1, 2, \ldots, n)$ (states), evolving towards some optimal solution/state $\boldsymbol{x}_*$. This is driven by the selection mechanism in an algorithm $A(p, t)$ with a set of parameter p, evolving with pseudotime t. In essence, an algorithm for minimization can also written schematically as

$$f(\boldsymbol{x}_i) \stackrel{A(p,t)}{\longrightarrow} f_{\min}(\boldsymbol{x}_*). \tag{4.9}$$

Furthermore, we can systematically compare the similarities and differences between self-organization and algorithms, which is summarized in Table 4.2.

Despite these similarities, there are some crucial differences between algorithms and self-organization. First, for self-organization, the exact avenue to self-organization may not be clear, but for algorithms, the ways of solution generations are often clear. Second, for self-organization, time is not an important factor per se, but for algorithms, the number of iterations (pseudotime) is crucially important because an effective algorithm should be able to find the optimal solutions using as least amount of computational efforts as possible. Third, the structure in self-organization is important, while the converged solutions themselves (not necessarily the structure) are most relevant in algorithms. Finally, the exact conditions of self-organization may be physically maintained, but for algorithms, the conditions for convergence can often lead to undesired premature convergence, and it is still not clear yet how to avoid such premature convergence in algorithms.

4.8 Markov Chain Monte Carlo

As we mentioned earlier, algorithms can be considered as biased Monte Carlo since the solutions generated by an algorithm is a statistical sampling method such as Monte Carlo [35]. In general, Monte Carlo methods are closely associated with

Markov chains. A Markov chain is a chain whose next state will depend only on the current state and the transition probability.

The solution sets generated by an algorithm essentially form a system of Markov chains and thus it is natural that the basic Markov chain theory can provide a generalized framework for analysing nature-inspired algorithms [48, 100, 105]. For example, Suzuki carried out a simple analysis of genetic algorithms using Markov chain theory [100], while He et al. used a discrete-time Markov chain approach and have proved that the flower pollination algorithm can have guaranteed global convergence under certain conditions [48].

At an even higher level, we can view algorithm systems as systems of multiple, interacting Markov chains that evolve with time. For example, a generalized approach has been designed using a Markov chain Monte Carlo method for global optimization [43]. In practice, this approach may converge slower than nature-inspired algorithms, and one of the reasons is that the selection mechanism is relatively weak in the generalized Markov chain model. Despite this drawback, this methodology can provide a quite general framework for optimization.

Mathematically speaking, Markov chain theory can provide some significant insight into algorithms. The largest eigenvalue of a proper Markov chain is unity, while the second largest eigenvalue λ_2 of the transition probability matrix essentially controls the rate of convergence of the Markov chain. However, it is very challenging to find this eigenvalue in practice. Even some estimates or approximations can be difficult. Therefore, the information and insight we can obtain is limited in practice, which may also limit its practical use.

4.8.1 Biased Monte Carlo

From the sampling point of view, nature-inspired algorithms share some similarity with the well-known Monte Carlo method [35]. In many algorithms, the initialization is done by random sampling of the search space, often using some uniformly distributed random numbers, and the initial population generated by such randomization is essentially the same as those by Monte Carlo. In addition, a solution vector can also be considered as a sampling point in the design space, and, in this sense, the set of solutions during iterations forms a sampling set.

However, there are some crucial differences. The samples generated by Monte Carlo sampling and its many variants tend to be distributed relatively uniformly in the design space and sometimes they are far away from each other in case of low discrepancy random numbers, while samples generated by nature-inspired algorithms will gradually aggregate towards some preferred regions based on the fitness of the solutions. Thus, the overall sampling process in algorithms is biased towards some promising regions where the optima and global optima may lie. The biased moves are guided by the fitness and the local information from the objective landscape. For example, the current best solution $\boldsymbol{g}^*$ in PSO acts as a local guide to attract biased moves. In this sense, we can consider all nature-inspired algorithms as information-guided biased Monte Carlo.

4.8.2 *Random Walks*

From probability theories, we know that the moves to generate solutions can be considered as random walks, modifying an existing solution $\boldsymbol{x}_N$ at step N by a perturbation $\boldsymbol{w}_N$. Mathematically speaking, a random walk can be written as

$$\boldsymbol{x}_{N+1} = \boldsymbol{x}_N + \boldsymbol{w}_N, \tag{4.10}$$

where $\boldsymbol{w}_N$ is a vector of random numbers (steps) drawn from a known probability solution. If $\boldsymbol{w}_N$ is drawn from a Gaussian distribution, then the random walks are isotropic. The movements in this case are often referred to as normal diffusion or Brownian motion. The expected distance moved (R) can be estimated by

$$R(N) \propto \sqrt{N}, \tag{4.11}$$

which has a square-root scaling property.

If the steps are drawn from a fat-tailed distribution such as Lévy distribution or Cauchy distribution, the diffusion becomes anomalous. In general, the above scaling property becomes

$$R(N) \propto N^q, \quad q > 0. \tag{4.12}$$

If $q \geq 1/2$, the diffusion is called superdiffusion [79]. Both Lévy distribution and Cauchy distribution for step sizes can have a fraction of large steps, which will lead to superdiffusion. This means that averaged distance increases faster than that for normal diffusion, which can potentially lead to a higher search efficiency if used properly in algorithms. For example, for Lévy flights, we have

$$q = \frac{3 - \lambda}{2}, \tag{4.13}$$

where $1 < \lambda \leq 2$ is the exponent of the power-law approximation to Lévy distribution

$$L(w) \sim |w|^{-1-\lambda}, \tag{4.14}$$

where $\sim$ denotes to draw random numbers from a distribution on the right-hand side. In fact, cuckoo search and flower pollination algorithm have used such fat-tailed Lévy flights for global search.

4.9 No-Free-Lunch Theorems

Though there are many algorithms in the literature, different algorithms can have different advantages and disadvantages. Thus, some algorithms are more suitable to solve certain types of problems than others. However, it is worth pointing out that there is no single algorithm that can be most efficient to solve all types of problems as dictated by the no-free-lunch (NFL) theorems [107]. Their rigorous proof requires some simplifications and assumptions. Two noticeable assumptions are (1) the set of points/solutions visited by an algorithm must be close under permutation, and (2) the points found in the iteration history are non-revisiting in subsequent iterations. In addition, the performance measure is based on the averaged performance over all possible functions and problems.

Let us provide here an informal way of looking at the no-free-lunch theorem from a slightly different perspective. For any univariate objective function $\phi(x)$ in a domain $[a, b]$, the mean of the function is

$$\mu = \frac{1}{b-a} \int_a^b \phi(x) dx. \tag{4.15}$$

Using the mean value theorem in calculus, we have

$$\frac{1}{b-a} \int_a^b \phi(x) dx = \phi(c), \quad c \in (a, b), \tag{4.16}$$

which suggests that the mean is a constant $\phi(c)$. If $\phi(x)$ can take any values and forms (including random values), we can treat $\phi(x)$ as a random variable. Then, $\mu = \phi(c)$ is the expectation. In a special case if all functions $\phi(x)$ are scaled to [0,1], then it can be expected that $\mu = 1/2$. Since μ is a constant, this means that the averaged objective landscape of all possible functions $\phi(x)$ becomes 'flat', which in turn means that there is no selection pressure for evolution of solutions. Consequently, it is no surprise that any algorithm (including a random search) can have equal efficiency. However, in practice, we do not need to solve all problems using a single algorithm.

Even the no-free-lunch theorems hold under certain conditions, these conditions may not be rigorously true for actual algorithms. For example, one condition for proving these theorems is the so-called non-revisiting condition. That is, the points during iterations form a path, and these points are distinct and will not be visited exactly again, though their nearby neighbourhood can be revisited. This condition is not strictly valid because almost all algorithms for continuous optimization will revisit some of their points in history. Such minor violation of assumptions can potentially leave room for free lunches. It has also been shown that under the right conditions such as co-evolution, certain algorithms can be more effective [108].

In addition, the work by T. Joyce and J.M. Heremann on the review of no-free-lunch (NFL) theorems [53] suggests that free lunches may exist for a finite set of problems, especially those algorithms that can exploit the objective landscape structure and knowledge of optimization problems to be solved. If the performance is not averaged over all *possible* problems, then free lunches can exist.

In fact, for a given finite set of problems and a finite set of algorithms, the comparison is essentially equivalent to a zero-sum ranking problem. In this case, some algorithms can perform better than others for solving a *certain type* of problems. In fact, almost all research papers published about comparison of algorithms use a few algorithms and a finite set (usually under 100 benchmarks), such comparison is essentially ranking. However, it is worth pointing out that for a finite set of benchmarks, the conclusions (e.g., ranking) obtained can only apply for that set of benchmarks, they may not be valid for other sets of benchmarks, and the conclusions can be significantly different for different sets of benchmarks. If interpreted in this sense, such comparison studies and their conclusions are consistent with NFL theorems.

Chapter 5
Mathematical Analysis of Algorithms: Part II

The perspectives of analysing algorithms we have given so far have been mainly following the mainstream literature on optimization, numerical analysis and operations research. Algorithms can also be analysed from other perspectives from other disciplines such as swarm intelligence, signal and image processing, machine learning, control theory and Bayesian framework.

5.1 Swarm Intelligence

Many new algorithms that are based on swarm intelligence (SI) may have drawn inspiration from different sources, but they have some similarities to some of the components that are used in particle swarm optimization (PSO) [57] and ant colony optimization (ACO) [31]. In this sense, PSO and ACO pioneered the basic ideas of swarm intelligence based computation [50, 61, 118].

From the algorithm point of view, the aim of a swarming system is to let the system evolve and converge into some stable states (ideally, some optimality). In this case, it has strong similarity to a self-organizing system. Such an iterative, self-organizing system can evolve, according to a set of rules or mathematical equations. As a result, such a complex system can interact and self-organize into certain converged states, showing some emergent characteristics of self-organization. Thus, the proper design of an efficient optimization algorithm is equivalent to finding efficient ways to mimic the evolution of a self-organizing system. In practice, all nature-inspired algorithms try to mimic some successful characteristics of biological, physical or chemical systems in nature.

Among all evolutionary algorithms, algorithms based on swarm intelligence dominate the landscape. There are many reasons for this dominance, though three obvious reasons are: (1) Swarm intelligence uses multiple agents as an evolving, interacting population, and thus provides good ways to mimic natural

X.-S. Yang, X.-S. He, *Mathematical Foundations of Nature-Inspired Algorithms*, SpringerBriefs in Optimization, https://doi.org/10.1007/978-3-030-16936-7_5

systems. (2) Population-based approaches allow parallelization and vectorization for implementations in practice, and are thus straightforward to implement. (3) These SI-based algorithms are usually flexible and yet sufficiently efficient to deal with a wide range of problems.

5.2 Filter Theory

In telecommunications and signal processing, signals are processed and filtered so as to gain certain desired properties [118, 128, 133]. If we consider the solutions during iterations are signals, the action of an algorithm is to filter out undesired signals (solutions) and let desired signals (good solutions including the best solution) pass through the system. As filtering occurs at multiple stages during iterations, the final well-filtered solutions are essentially the converged solution set.

In this sense, the design of algorithms is equivalent to the design of filters. Therefore, for linear algorithms such as PSO and bat algorithm, it is possible to use filter theory and signal processing techniques to analyse them.

This point of view of looking at algorithms may give different insights into the algorithms and their functionalities. Though we have not seen such studies in the literature, it is no doubt future research will investigate this route further.

5.3 Bayesian Framework and Statistical Analysis

All nature-inspired algorithms use some form of initial configurations that are often randomly initialized using a uniform distribution between some fixed ranges or bounds. As the iterations continue, the actual distribution of the population or solutions may be significantly different from the initial uniform distributions. In fact, a good algorithm should be able to 'forget' its initial states such that the solution set can be viewed as a set of interacting Markov chains. This way the algorithm can converge to some desired or converged states. Among these converged states, the optimal solutions should exist, or likely to exist as the number of iterations continues to increase.

From the perspective of the Bayesian framework, the posterior probability of finding the global optimal solution should gradually increase, even starting from a uniform prior distribution. Ideally, at each iteration k with new information (I_k, or data), the probability $P(G_k|I_k)$ should improve, given the prior probability $P(G_k)$. Mathematically, this is equivalent to

$$P(G_k|I_k) = \frac{P(I_k|G_k)P(G_k)}{P(I_k)}, \tag{5.1}$$

where $P(I_k)$ is the marginal likelihood or scaling factor. Here, G_k can be considered loosely as a hypothesis for the algorithm to find the global optimal solution, which can be affected by the new data or solution sets obtained during iterations.

The initial spread or variation of the probability distribution is large. As the iterations proceed, the posterior probability usually becomes narrower, focusing more on some regions or subsets of the search space. If the focused region happens to contain the global optimal solution, the algorithm is likely to find it. However, if the focused region does not contain the global optimality, there is no guarantee to find the global optimal solution. Thus, the ways to modify the posterior probability, based on new data, are essential to design efficient algorithms. However, how to achieve such modifications effectively is still an open question.

In most cases, it may be difficult to show what kind of posterior probability distribution may become at each iteration. However, the diversity of the solutions can be measured by the population variance. For example, Zaharie carried out a variance analysis of population and the effect of mutation on solution diversity in differential evolution [136]. The variance of the population $\mathrm{var}(P_t)$ at time t is governed by

$$\mathrm{var}(P_t) = Q(F, p_m, t)\, \mathrm{var}(P_0), \tag{5.2}$$

where F is a constant and p_m is the effective mutation probability. This relationship links the variance of the current population P_t of n solutions with that of the initial population P_0 [136]. In addition, we have

$$Q(F, p_m, t) = \Big[1 - 2F^2 p_m - \frac{p_m(2 - p_m)}{n}\Big]^t, \tag{5.3}$$

which defines a critical value of F when $Q = 1$.

Indeed, variance analysis provides some information about the diversity of the population during the iterations. However, this kind of analysis requires that the system is linear, and thus it cannot directly be extended to analyse nonlinear systems such as the firefly algorithm.

5.4 Stochastic Learning

Though one of the intention of the nature-inspired algorithms is to be flexible and problem-independent, this means that these algorithms often treat optimization problems as a black box without using any problem-specific information. Such black-box approaches may be easy to implement, but they lack the ability to learn from experience or data.

On the other hand, many machine learning algorithms use data and latest information extensively. This is especially true for artificial neural networks, support vector machines and many other learning algorithms [76, 103, 104, 118]. In a broader sense, an algorithm can be considered as a 'learning' iterative procedure where new information such as gradients and the errors between the predictions and targets acts as new data to train the algorithm and steer the ways or paths in the search space. From this point of view, algorithms are stochastic learning algorithms to a different extent. However, the learning abilities of different algorithms (in terms of using data and new solutions) can be hugely different, from the strong ones such as neural networks and gradient-based algorithms to the weakest (or almost none) purely random search algorithms.

In fact, a good trend is to somehow incorporate some problem-specific knowledge, learned on the go, so as to enhance the performance of an algorithm used. For some computational extensive and expensive applications, some good and simple approximations to the objective surface can significantly reduce the computational costs. Such surrogate-based optimization can be very useful to real-world applications in aerospace engineering, computational biology, bioinformatics and large-scale simulations.

It can be expected that techniques inspired by machine learning can be useful to tune optimization algorithms and design better algorithms in the future, which will in turn be applied back to data mining and machine learning to solve important problems more effectively.

5.5 Parameter Tuning and Control

Almost all nature-inspired algorithms have algorithm-dependent parameters and the actual values of these parameters can affect the overall performance of an algorithm significantly. Some parameters may have a strong influence, while others may have a weaker influence.

From a mathematical point of view, an algorithm A tends to generate a new and better solution $\boldsymbol{x}^{t+1}$ to a given problem from the current solution $\boldsymbol{x}^t$ at iteration or time t. In modern metaheuristic algorithms, randomization is often used in an algorithm, and in many cases, randomization appears in the form of a set of m random variables $\boldsymbol{\varepsilon} = (\varepsilon_1, \ldots, \varepsilon_m)$ in an algorithm. For example, in simulated annealing, there is one random variable, while in particle swarm optimization, there are two random variables. In addition, there are often a set of k parameters in an algorithm. For example, in particle swarm optimization, there are four parameters (two learning parameters, one inertia weight and the population size). In general, we can have a vector of parameters $\boldsymbol{p} = (p_1, \ldots, p_k)$. Mathematically speaking, we can write an algorithm with k parameters and m random variables as

$$\boldsymbol{x}^{t+1} = \boldsymbol{A}\Big(\boldsymbol{x}^t, \boldsymbol{p}(t), \boldsymbol{\varepsilon}(t)\Big), \tag{5.4}$$

where $\boldsymbol{A}$ is a nonlinear mapping from a given solution (a D-dimensional vector $\boldsymbol{x}^t$) to a new solution vector $\boldsymbol{x}^{t+1}$.

Representation (5.4) gives rise to two types of optimality: optimality of a problem and optimality of an algorithm. For an optimization problem such as min $f(\boldsymbol{x})$, there is a global optimal solution whatever the algorithmic tool we may use to find this optimality. This is the optimality for the optimization problem. On the other hand, for a given problem Φ with an objective function $f(\boldsymbol{x})$, there are many algorithms that can solve it. Some algorithms may require less computational effort than others. There may be the best algorithm with the least computing cost, though this may not be unique. However, this is not our concern here. Once we have chosen an algorithm A to solve a problem Φ, there is an optimal parameter setting for this algorithm so that it can achieve the best performance. This optimality depends on both the algorithm itself and the problem it solves. We will focus on this type of optimality.

That is, the optimality to be achieved is

$$\text{Maximize the performance of } \xi = A(\Phi, \boldsymbol{p}, \boldsymbol{\varepsilon}), \tag{5.5}$$

for a given problem Φ and a chosen algorithm $A(., \boldsymbol{p}, \boldsymbol{\varepsilon})$. We will denote this optimality as $\xi_* = A_*(\Phi, \boldsymbol{p}_*) = \xi(\Phi, \boldsymbol{p}_*)$, where $\boldsymbol{p}_*$ is the optimal parameter setting for this algorithm so that its performance is the best [130]. Here, we have used a fact that $\boldsymbol{\varepsilon}$ is a random vector that can be drawn from some known probability distributions; thus, the randomness vector should not be related to the algorithm optimality.

It is worth pointing out that there is another potential optimality. That is, for a given problem, a chosen algorithm with the best parameter setting $\boldsymbol{p}_*$, we can still use different random numbers drawn from various probability distributions and even chaotic maps, so that even better performance may be achieved. Strictly speaking, if an algorithm $A(., ., \boldsymbol{\varepsilon})$ has a random vector $\boldsymbol{\varepsilon}$ that is drawn from a uniform distribution $\boldsymbol{\varepsilon}_1 \sim U(0, 1)$ or from a Gaussian $\boldsymbol{\varepsilon}_2 \sim N(0, 1)$, it becomes two algorithms $A_1 = A(., ., \boldsymbol{\varepsilon}_1)$ and $A_2 = A(., ., \boldsymbol{\varepsilon}_2)$. Technically speaking, we should loosely treat them as different algorithms. Since our emphasis here is about parameter tuning so as to find the optimal setting of parameters, we will omit the effect of randomness vectors, and thus focus on

$$\text{Maximize } \xi = A(\Phi, \boldsymbol{p}). \tag{5.6}$$

In essence, tuning algorithm involves the fine-tuning of its algorithm-dependent parameters. Therefore, parameter tuning is equivalent to algorithm tuning in the present context.

5.5.1 Parameter Tuning

In order to tune $A(\Phi, \boldsymbol{p})$ so as to achieve its best performance, a parameter-tuning tool, i.e., a tuner, is needed. Like tuning a high-precision machinery, sophisticated tools are required. For tuning parameters in an algorithm, what tool can we use?

One way is to use a better, existing tool (say, algorithm B) to tune an algorithm A. Now the question may become: how do you know B is better? Is B well-tuned? If yes, how do you tune B in the first place? Naively, if we say, we use another tool (say, algorithm C) to tune B. Now again the question becomes how algorithm C has been tuned? This can go on and on, until the end of a long chain, say, algorithm Q. In the end, we need some tool/algorithm to tune this Q, which again comes back to the original question: how to tune an algorithm A so that it can perform best?

It is worth pointing out that even if we have good tools to tune an algorithm, the best parameter setting and thus performance all depend on the performance measures used in the tuning. Ideally, these parameters should be robust enough to minor parameter changes, random seeds and even problem instance. However, in practice, they may not be achievable. According to Eiben [32], parameter tuning can be divided into iterative and non-iterative tuners, single-stage and multistage tuners. The meaning of these terminologies is self-explanatory. In terms of the actual tuning methods, existing methods include sampling methods, screening methods, model-based methods and metaheuristic methods. Their success and effectiveness can vary, and thus there are no well-established methods for universal parameter tuning. One promising approach for parameter tuning is the self-tuning framework in combination with an optimization algorithm [130], which used the firefly algorithm as an example.

5.5.2 Parameter Control

Parameter tuning is just one important aspect of the algorithm tuning. In parameter tuning, once an algorithm is tuned, the values of parameters are fixed, and their values will not change with the iteration counter k or t. However, there is no strong evidence that such fixed settings are best. If we allow the parameters to vary with iteration, this becomes parameter control. In fact, studies show that parameter control can be advantageous.

For example, in the standard differential evolution (DE), the parameter F is fixed in the range of 0–1, often between 0.5 and 0.9 in many implementations. In the adaptive different evolution, this parameter varies randomly between 0.1 and 0.9.

Similarly, in the standard firefly algorithm (FA), the attractiveness parameter β_0 is fixed as $\beta_0 = 1$ or $\beta_0 = 0.5$. Recent studies showed that a uniformly distribution between 0.2 and 1 can enhance the performance of the algorithm.

In addition, many variants, especially those using chaos, use some known distributions or iterative chaotic maps to replace a fixed parameter, and many studies in the literature have demonstrated that such chaotic variants can usually have better mixing abilities and the ability of escaping from the local optima. For example, Gandomi and Yang [40] used chaotic maps to design a chaotic bat algorithm, which indeed enhanced the overall performance for many applications.

5.6 Hyper-Optimization

From our earlier observations and discussions, it is clear that parameter tuning is the process of optimizing the optimization algorithm; therefore, it is a hyper-optimization problem. In essence, a tuner is a meta-optimization tool for tuning algorithms.

For a standard unconstrained optimization problem, the aim is to find the global minimum f_* of a function $f(\boldsymbol{x})$ in a D-dimensional space. That is,

$$\text{Minimize } f(\boldsymbol{x}), \quad \boldsymbol{x} = (x_1, x_2, \ldots, x_D)^T \in \mathbb{R}^D. \tag{5.7}$$

Once we choose an algorithm A to solve this optimization problem, the algorithm will find a minimum solution $f_{\min}$ which may be close to the true global minimum f_*. For a given tolerance δ, this may require t_δ iterations to achieve $|f_{\min} - f_*| \leq \delta$. Obviously, the actual t_δ will largely depend on both the problem objective $f(\boldsymbol{x})$ and the parameters $\boldsymbol{p}$ of the algorithm used.

The main aim of algorithm tuning is to find the best parameter setting $\boldsymbol{p}_*$ so that the computational cost or the number of iterations t_δ is the minimum. Thus, parameter tuning as a hyper-optimization problem can be written as

$$\text{Minimize } t_\delta = A(f(\boldsymbol{x}), \boldsymbol{p}), \tag{5.8}$$

whose optimality is $\boldsymbol{p}_*$.

Ideally, the parameter vector $\boldsymbol{p}_*$ should be sufficiently robust. For different types of problems, any slight variation in $\boldsymbol{p}_*$ should not affect the performance of A much, which means that $\boldsymbol{p}_*$ should lie in a flat range, rather than at a sharp peak in the parameter landscape.

5.6.1 A Multiobjective View

If we look the algorithm-tuning process from a different perspective, it is possible to construct it as a multiobjective optimization problem with two objectives: one objective $f(\boldsymbol{x})$ for the problem Φ and one objective t_δ for the algorithm. That is,

$$\text{Minimize } f(\boldsymbol{x}) \text{ and Minimize } t_\delta = A(f(\boldsymbol{x}), \boldsymbol{p}), \tag{5.9}$$

where t_δ is the (average) number of iterations needed to achieve a given tolerance δ so that the found minimum $f_{\min}$ is close enough to the true global minimum f_*, satisfying $|f_{\min} - f_*| \leq \delta$.

This means that for a given tolerance δ, there will be a set of best parameter settings with a minimum t_δ. As a result, the bi-objectives will form a Pareto front. In principle, this bi-objective optimization problem (5.9) can be solved by any methods that are suitable for multiobjective optimization. But as δ is usually given, a natural way to solve this problem is to use the so-called ϵ-constraint or δ-constraint methods. The naming may be dependent on the notations; however, we will use δ-constraints.

For a given $\delta \geq 0$, we change one of the objectives (i.e., $f(\boldsymbol{x})$) into a constraint, and thus the above problem (5.9) becomes a single-objective optimization problem with a constraint. That is,

$$\text{Minimize } t_\delta = A(f(\boldsymbol{x}), \boldsymbol{p}), \tag{5.10}$$

subject to

$$f(\boldsymbol{x}) \leq \delta. \tag{5.11}$$

In the rest of this chapter, we will set $\delta = 10^{-5}$.

The important thing is that we still need an algorithm to solve this optimization problem. However, the main difference from a common single-objective problem is that the present problem contains an algorithm A. Ideally, an algorithm should be independent of the problem, which treats the objective to be solved as a black box. Thus we have $A(., \boldsymbol{p}, \boldsymbol{\varepsilon})$; however, in reality, an algorithm will be used to solve a particular problem Φ with an objective $f(\boldsymbol{x})$. Therefore, both notations $A(., \boldsymbol{p})$ and $A(f(\boldsymbol{x}), \boldsymbol{p})$ will be used here.

5.6.2 Self-tuning Framework

This framework has been proposed by Yang et al. in 2013 [130]. In principle, we can solve (5.10) by any efficient or well-tuned algorithm. Now a natural question is: can we solve this algorithm-tuning problem by algorithm A itself? There is no reason why we cannot. In fact, if we solve (5.10) by using A, we have a self-tuning algorithm. That is, the algorithm automatically tunes itself for a given problem objective to be optimized. This essentially provides a framework for a self-tuning algorithm as shown in Figure 5.1.

This framework is generic in the sense that any algorithm can be tuned this way, and any problem can be solved within this framework. This essentially achieves two goals simultaneously: parameter tuning and optimality finding.

In the rest of this section, we will use firefly algorithm (FA) as a case study to self-tune FA for a set of function optimization problems.

Implement an algorithm $A(., \boldsymbol{p}, \boldsymbol{\varepsilon})$
 with parameters $\boldsymbol{p} = [p_1, ..., p_K]$ and random vector $\boldsymbol{\varepsilon} = [\varepsilon_1, ..., \varepsilon_m]$;
Define a tolerance (e.g., $\delta = 10^{-5}$);
 Algorithm objective $t_\delta(f(\boldsymbol{x}), \boldsymbol{p}, \boldsymbol{\varepsilon})$;
 Problem objective function $f(\boldsymbol{x})$;
 Find the optimality solution $f_{\min}$ within δ;
 Output the number of iterations t_δ needed to find $f_{\min}$;
 Solve $\min t_\delta(f(\boldsymbol{x}), \boldsymbol{p})$ using $A(., \boldsymbol{p}, \boldsymbol{\varepsilon})$ to get the best parameters;
Output the tuned algorithm with the best parameter setting $\boldsymbol{p}_*$.

Fig. 5.1 A framework for a self-tuning algorithm

5.6.3 Self-tuning Firefly Algorithm

Now let us use the framework outlined earlier to tune the firefly algorithm (FA). As we have seen earlier in Chapter 2, FA has the following updating equation:

$$x_i^{t+1} = x_i^t + \beta_0 e^{-\gamma r_{ij}^2}(x_j^t - x_i^t) + \alpha\, \epsilon_i^t, \tag{5.12}$$

which contains four parameters: α, β_0, γ and the population size n. For simplicity for parameter tuning, we set $\beta_0 = 1$ and $n = 20$, and therefore, the two parameters to be tuned are: $\gamma > 0$ and $\alpha > 0$. It is worth pointing out that γ controls the scaling, while α controls the randomness. For this algorithm to convergence properly, randomness should be gradually reduced, and one way to achieve such randomness reduction is to use

$$\alpha = \alpha_0 \theta^t, \quad \theta \in (0, 1), \tag{5.13}$$

where t is the index of iterations/generations. Here α_0 is the initial randomness factor, and we can set $\alpha_0 = 1$ without losing generality. Therefore, the two parameters to be tuned become γ and θ. For the detailed procedure and results, readers can refer to [130].

Our framework for self-tuning algorithms is truly self-tuning in the sense that the algorithm to be tuned is used to tune itself. We have used the firefly algorithm and a set of test functions to test the proposed self-tuning algorithm framework. The results have shown that it can indeed work well [130]. We also found that some parameters require fine-tuning, while others do not need to be tuned carefully. This is because different parameters may have different sensitivities, and thus may affect the performance of an algorithm in different ways. Only parameters with high sensitivities need careful tuning.

Though this self-tuning framework is successful, it requires further extensive testing with a variety of test functions and many different algorithms. It may also be possible to see how probability distributions can affect the tuned parameters and even the parameter-tuning process. In addition, it can be expected that this present framework is also useful for parameter control, so a more generalized framework for both parameter tuning and control can be used for a wide range of applications. Furthermore, our current framework may be extended to multiobjective problems so that algorithms for multiobjective optimization can be tuned in a similar way.

5.7 Multidisciplinary Perspectives

The above analyses clearly indicate that a particular method for analysis can usually look at the algorithms from one perspective at a time. Different approaches and perspectives can provide different insights, potentially complementary to each other. Therefore, multidisciplinary approaches are needed to analyse algorithms from multiple angles so as to provide a fuller picture, including convergence analysis, stability analysis, sensitivity analysis, and robustness analysis as well as parallelism in implementation.

It can be expected that a multidisciplinary framework can be formulated to analyse algorithms comprehensively, and such a framework requires all the above disciplines and methodologies to work together. It is hoped that this work can inspire more research in this area.

5.8 Future Directions

The research area of nature-inspired algorithms is very active, and there are many hot topics for further research directions concerning these algorithms. Here we highlight a few:

- **Theoretical Framework**: Though there are many studies concerning the implementations and applications of metaheuristic algorithms, mathematical analysis of such algorithms lags behind. Though we have tried to analyse these algorithms both qualitatively and quantitatively from different angles and perspectives, the work is far from complete. There is a strong need to build a unified, theoretical framework to analyse these algorithms mathematically so as to gain further in-depth understanding in an even more rigorous manner.

  In addition, many open problems remain unresolved. For example, it is not clear how local rules can lead to the rise of self-organized structure in algorithms. More generally, it still lacks key understanding about the rise of collective intelligence and swarm intelligence in a multi-agent system and their exact conditions.
- **Parameter Tuning**: Almost all algorithms have algorithm-dependent parameters, and the performance of an algorithm is largely influenced by its parameter setting. As we have seen above, some parameters may have stronger influence

than others. Therefore, proper parameter tuning and sensitivity analysis are needed to tune algorithms to their best. However, such tuning is largely done by trial and error in combination with the empirical observations. Even we have established a self-tuning framework, it can still be time-consuming to tune new algorithms. How to tune them quickly and automatically is still an open question.

- **Sampling and Initialization**: Empirical observations and simulations suggest that the initial configuration of the evolving population can have a heavy influence on the final results for some algorithms. Ideally, an efficient algorithm should be independent of its initial settings, though it is difficult to achieve in practice. It is still unclear how to understand the exact role of initialization. Most algorithms use uniform distributions for initialization, which can be considered to be consistent with most Monte Carlo approaches.

  In principle, the effect of initialization and sampling of the initial solutions in the search space can be loosely understood in terms of Bayesian statistics. However, how to analyse it rigorously is still an open question. In addition, different sampling techniques may also have different influence on the algorithmic behaviour. For example, initialization for higher-dimensional problems using Latin hypercube sampling may be different from the Gibbs sampler for some problems and algorithms. Some systematic studies and comparison are needed to gain better insight concerning this issue.
- **Algorithmic Structure**: We know different algorithms may have different structures in updating solutions and generating new solutions, and such algorithmic structures may have a strong influence on the performance of an algorithm. However, it is not clear how the algorithmic structure affects the performance in a particular algorithm and what structural features or structures may be beneficial to design better algorithms.

  For hybrid algorithms, the proper combination of different structural features can also be very important, though it is not yet fully understood how to achieve a good combination of different algorithms so as to produce better hybrids.
- **Large-Scale Applications**: Despite the success of nature-inspired algorithms and their diverse applications, most studies in the literature have concerned problems of moderate sizes with the number of variables up to a few hundred at most. In real-world applications, there may be thousands and even millions of design variables, it is not clear yet how these algorithms can be scaled up to solve such large-scale problems.

  In addition, though these algorithms have been applied to solve combinatorial problems such as scheduling and the travelling salesman problem with promising results, these problems are typically non-deterministic polynomial-time (NP) hard, and, thus for larger problem sizes, they can be very challenging to solve. Researchers are not sure how to modify existing algorithms to cope with such challenges.
- **Hybrid and Co-evolutionary Algorithms**: The algorithms we have covered here are algorithms that are 'pure' and 'standard' in the sense that they have not been heavily modified by hybridizing with others. Both empirical observations and studies show that the combination of the advantages from two or more

different algorithms can produce a better hybrid, which can use the distinct advantages of its component algorithms and potentially avoid their drawbacks. In addition, it is possible to build a proper algorithm system, or algorithm-generating systems, to allow a few algorithms to co-evolve to obtain an overall better performance.

Though NFL theorems may hold for simple algorithms, it has been shown that there can be free lunches for co-evolutionary algorithms [108]. Therefore, future research can focus on figuring out how to assemble different algorithms into an efficient co-evolutionary system and then tune the system to its best.

- **Self-adaptive and Self-evolving Algorithms**: Sometimes, the parameters in an algorithm can vary to suit for different types of problems. Ideally, an algorithm should be self-adaptive and be able to automatically tune itself to suit for a given type of problems without much supervision from the users. Such algorithms should also be able to evolve by learning from their past performance histories. The ultimate aim for researchers is to build a set of self-adaptive, self-tuning, self-learning and self-evolving algorithms that can solve a diverse range of real-world applications efficiently and quickly in practice.

Nature-inspired algorithms have become a powerful tool set for solving optimization problems and their studies form an active area of research. Though we have attempted to summarize different ways of analysing algorithms and to highlight the main issues, there are still many open problems that need to be resolved in the future. It is hoped that this work can inspire more research concerning the above important topics.

Chapter 6
Applications of Nature-Inspired Algorithms

Nature-inspired algorithms have become powerful and popular for solving problems in optimization, computational intelligence, data mining, machine learning, transport and vehicle routing. After the theoretical analyses in earlier chapters, it would be useful to provide examples and case studies to show that these algorithms can indeed work well in practice [52, 118, 120, 128].

The diversity of the applications of these nature-inspired optimization algorithms is vast and the literature is rapidly expanding. Thus, it is not possible to cover even a small fraction of the recent applications and large-scale applications [19, 120].

Therefore, this chapter only attempts to provide a snapshot of ten areas of applications with a brief review of relevant literature.

6.1 Design Optimization in Engineering

One of the main areas of applying nature-inspired algorithms is to solve design optimization problems in engineering and industries. There are thousands of papers published each year concerning the optimal designs of key components, products and systems. Here, we only briefly touch a few benchmarks.

6.1.1 Design of a Spring

Let us start with a simple but nonlinear problem about the design of a spring under tension or compression from a metal wire [41, 126]. There are three design variables: the wire diameter (r), the mean coil diameter (d) and the number (N) of turns/coils. The objective is to minimize the overall weight of the spring

$$\text{minimize} \quad f(\boldsymbol{x}) = (2 + N)r^2 d, \tag{6.1}$$

X.-S. Yang, X.-S. He, *Mathematical Foundations of Nature-Inspired Algorithms*,
SpringerBriefs in Optimization, https://doi.org/10.1007/978-3-030-16936-7_6

Table 6.1 Comparison of optimal solutions for spring design

Author	Optimal solution	Best objective
Arora [7]	(0.053396, 0.399180, 9.185400)	0.01273
Coello [26]	(0.051480, 0.351661, 11.632201)	0.01271
Yang and Deb [122]	(0.051690, 0.356750, 11.28716)	0.012665
Yang et al. [134]	(0.051690, 0.356750, 11.28716)	0.012665

subject to nonlinear constraints:

$$g_1(\boldsymbol{x}) = 1 - \frac{Nd^3}{71785r^4} \le 0, \quad g_2(\boldsymbol{x}) = \frac{d(4d-r)}{12566r^3(d-r)} + \frac{1}{5108r^2} - 1 \le 0, \tag{6.2}$$

$$g_3(\boldsymbol{x}) = 1 - \frac{140.45r}{d^2N} \le 0, \quad g_4(\boldsymbol{x}) = (d+r) - 1.5 \le 0. \tag{6.3}$$

Some simple bounds or limits for the design variables are

$$0.05 \le r \le 2.0, \quad 0.25 \le d \le 1.3, \quad 2.0 \le N \le 15.0. \tag{6.4}$$

We have solved this problem using the recently developed multi-species cuckoo search (MSCS) algorithm [134], and the results of 20 different runs are summarized in Table 6.1 where comparison is also made. As we can see, MSCS can obtain the best or the same results as the best results in the literature.

6.1.2 Pressure Vessel Design

A well-known design benchmark is the pressure vessel design problem that has been used by many researchers. The design objective is to minimize the overall cost of a cylindrical vessel subject to stress and volume requirements. There are four design variables: the thickness d_1 and d_2 for the head and body, respectively, the inner radius r, and the length W of the cylindrical section [20, 41]. The main objective is

$$\min f(\boldsymbol{x}) = 0.6224rWd_1 + 1.7781r^2d_2 + 19.64rd_1^2 + 3.1661Wd_1^2, \tag{6.5}$$

subject to four constraints:

$$g_1(\boldsymbol{x}) = -d_1 + 0.0193r \le 0, \quad g_2(\boldsymbol{x}) = -d_2 + 0.00954r \le 0, \tag{6.6}$$

$$g_3(\boldsymbol{x}) = -\frac{4\pi r^3}{3} - \pi r^2 W - 1296000 \le 0, \quad g_4(\boldsymbol{x}) = W - 240 \le 0. \tag{6.7}$$

The inner radius and length are limited to $10.0 \le r, W \le 200.0$. However, the thickness d_1 and d_2 can only be the integer multiples of a basic thickness of

Table 6.2 Comparison of optimal solutions for pressure vessel design

Author	Optimal solution	Best objective
Cagnina et al. [20]	(0.8125, 0.4375, 42.0984, 176.6366)	6059.714
Coello [26]	(0.8125, 0.4375, 42.3239, 200.0)	6288.7445
Yang et al. [129]	(0.8125, 0.4375, 42.0984456, 176.6365959)	6059.714
Yang et al. [134]	(0.8125, 0.4375, 42.0984456, 176.6366)	6059.714

0.0625 in. Thus, the simple bounds for thickness are

$$1 \times 0.0625 \le d_1, d_2 \le 99 \times 0.0625. \tag{6.8}$$

With these additional constraints, this optimization problem becomes a mixed integer programming because two variables are discrete and the other two variables are continuous.

Using the same MSCS algorithm, the results of 20 independent runs are summarized and compared in Table 6.2 [134]. In fact, all these algorithms are able to find the true global optimal solution with $f_{\min} = 6059.714$, as obtained analytically by Yang et al. [129].

6.1.3 Speed Reducer Design

The speed reducer design is an engineering design benchmark with seven design variables. These design variables include the face width of the gear, number of teeth, and diameter of the shaft and others [20, 41]. All these variables can take continuous values, except for x_3 which is an integer.

The objective is to minimize the cost function

$$f(\boldsymbol{x}) = 0.7854\Big[x_1 x_2^2 (3.3333 x_3^2 + 14.9334 x_3 - 43.0934) + (x_4 x_6^2 + x_5 x_7^2)\Big]$$
$$- 1.508 x_1 (x_6^2 + x_7^2) + 7.4777 (x_6^3 + x_7^3), \tag{6.9}$$

subject to 11 constraints:

$$g_1(\boldsymbol{x}) = \frac{27}{x_1 x_2^2 x_3} - 1 \le 0, \quad g_2(\boldsymbol{x}) = \frac{397.5}{x_1 x_2^2 x_3^2} - 1 \le 0, \tag{6.10}$$

$$g_3(\boldsymbol{x}) = \frac{1.93 x_4^3}{x_2 x_3 x_6^4} - 1 \le 0, \quad g_4(\boldsymbol{x}) = \frac{1.93 x_5^3}{x_2 x_3 x_7^4} - 1 \le 0, \tag{6.11}$$

$$g_5(\boldsymbol{x}) = \frac{1.0}{110 x_6^3} \sqrt{(\frac{745.0 x_4}{x_2 x_3})^2 + 16.9 \times 10^6} - 1 \le 0, \tag{6.12}$$

Table 6.3 Comparison of optimal solutions for the speed reducer problem

Author	Optimal solution	Best objective
Cagnita et al. [20]	(3.5, 0.7, 17, 7.3, 7.8, 3.350214, 5.286683)	2996.348165
Yang and Gandomi [126]	(3.5, 0.7, 17, 7.3, 7.71532, 3.35021, 5.2875)	2994.467
Yang et al. [134]	(3.5, 0.7, 17, 7.3, 7.8, 3.34336449, 5.285351)	2993.749589

$$g_6(\boldsymbol{x}) = \frac{1.0}{85x_7^3}\sqrt{(\frac{745.0x_5}{x_2x_3})^2 + 157.5 \times 10^6} - 1 \leq 0, \tag{6.13}$$

$$g_7(\boldsymbol{x}) = x_2x_3 - 40 \leq 0, \quad g_8(\boldsymbol{x}) = 5x_2 - x_1 \leq 0, \tag{6.14}$$

$$g_9(\boldsymbol{x}) = x_1 - 12x_2 \leq 0, \quad g_{10}(\boldsymbol{x}) = (1.5x_6 + 1.9) - x_4 \leq 0, \tag{6.15}$$

$$g_{11}(\boldsymbol{x}) = (1.1x_7 + 1.9) - x_5 \leq 0. \tag{6.16}$$

In addition, the simple bounds for the variables are: $2.6 \leq x_1 \leq 3.6$, $0.7 \leq x_2 \leq 0.8$, $17 \leq x_3 \leq 28$ (integers only), $7.3 \leq x_4 \leq 8.3$, $7.8 \leq x_5 \leq 8.4$, $2.9 \leq x_6 \leq 3.9$, and $5.0 \leq x_7 \leq 5.5$.

The results of 20 independent runs are summarized in Table 6.3 where comparison has also been made. As we can see, MSCS obtained the best results [134]. Since there is no literature about the analysis of this problem, no one knows what the global best solution should be. Thus, we can only say that 2993.749589 is the best result achieved so far.

6.1.4 Other Design Problems

Obviously, there are many other design problems and applications. For example, reliability-based design optimization concerns the designs subject to stochastic or probabilistic constraints due to the intrinsic uncertainty in materials properties and design requirements. Recent studies by Chakri et al. [22, 23] show that the directional bat algorithm (DBA) can obtain optimal design solutions effectively. As the focus of this book is not about applications, we will not discuss reliability-based designs any further.

6.2 Inverse Problems and Parameter Identification

Many problems in applications can be considered as inverse problems so as to estimate or identify the key parameter values for a given set of observations under a given configuration of a physical, chemical or biological system.

Table 6.4 Measured response of a simple vibration system

t	0.00	0.20	0.40	0.60	0.80	1.00	1.20	1.40	1.60	1.80	2.00
$y_d(t)$	0.00	0.59	1.62	2.21	1.89	0.69	−0.99	−2.53	−3.36	−3.15	−1.92

To show how to solve such type of problems, let us start with a simple example. For a simple vibration problem, two unknown parameters μ and ν are estimated from the measurements of vibration amplitudes. The governing equation is

$$\frac{d^2y(t)}{dt^2} + \mu\frac{dy(t)}{dt} + \nu y(t) = 40\cos(3t). \tag{6.17}$$

This is a damped harmonic motion problem and its general solution can be quite complex [119]. However, for fixed $\mu = 4$ and $\nu = 5$ with initial values of $y(0) = 0$ and $y'(0) = 0$, its analytical solution can be simplified as

$$y(t) = e^{-2t}[\cos(t) - 7\sin(t)] + 3\sin(3t) - \cos(3t). \tag{6.18}$$

For a real system with a forcing term $40\cos(3t)$, we do not know the parameters, but its vibrations can be measured. For example, in an experiment, there are $N = 11$ measurements as shown in Table 6.4.

The task is to estimate the values of the two parameters. However, one of the challenges is that the calculation of the objective function that is defined as the sum of errors squared. That is,

$$f(\boldsymbol{x}) = \sum_{i=1}^{N}(y_{i,\text{predicted}} - y_{i,d})^2, \tag{6.19}$$

where the predicted $y(t)$ has to be obtained by solving the second-order ordinary differential equation (6.17) numerically and iteratively for every given set of μ and ν. This becomes a series of optimization problems.

Using the MSCS with a population of 40 cuckoos in total [134], we have run the algorithm for 20 times so as to obtain meaningful statistics. The mean values of the two parameters obtained from the measured data are: $\mu_* = 4.025$ and $\nu_* = 4.981$, which are very close to the true values of $\mu = 4.000$ and $\nu = 5.000$.

6.3 Image Processing

Signal and image processing is an important area with many applications, from speech recognition to natural language processing, and from image segmentation to facial expression recognition in machine learning and artificial intelligence. The relevant literature in this area is vast.

In the area of crop classification from satellite images, Senthilnath et al. [91, 92] showed that the bat algorithm can be used to effectively identify crop types from multispectral satellite images. In addition, multi-temporal satellite image registration has been carried out by using nature-inspired algorithms, and Senthilnath et al. concluded that the firefly algorithm provided the best results in the most efficient way [91]. Furthermore, for the purpose of hyperspectral band selection of hyperspectral satellite images, the bat algorithm has been found to be the most effective [70], in combination with the optimal-path forest classifier.

6.4 Classification, Clustering and Feature Selection

Classification and clustering methods are important with many real-world applications. As a simple example, let us use the multi-species cuckoo search to solve the classification problem of the well-known Fisher's Iris flower data set. This data set has 150 data points or instances with 4 attributes and 3 distinct classes. We use the data from the UCI Machine Learning Repository.[1] The best accuracy we obtained is 97.1% [134], which is better than the results obtained by other methods.

For clustering, many researchers used nature-inspired algorithms to optimize the centroid locations for clustering, in combination with the traditional clustering methods such as the K-means and fuzzy K-means methods. They usually obtained much better results than traditional methods alone.

On the other hand, for feature selection, a binary cuckoo search has shown to be effective in selecting the most appropriate features among many attributes for test benchmark data sets such as diabetes, DNA, heart diseases and mushroom types [84]. Other studies also showed that all swarm intelligence based algorithms can produce competitive results [38].

6.5 Travelling Salesman Problem

The travelling salesman problem (TSP) concerns the tour of n cities once and exactly once starting from a city and returning to the same starting city so that the total distance travelled is the minimum.

There are many different ways to formulate the TSP, and there is a vast literature on this topic. Here, we will use the binary integer LP formulation by G. Dantzig.

Let $i = 1, 2, \ldots, n$ be n cities, which can be considered as the nodes of a graph. Let x_{ij} be the decision variable for connecting city i to city j (i.e., an edge of the graph from node i to node j) such that $x_{ij} = 1$ means the tour starts from i and ends

[1] http://archive.ics.uci.edu/ml/datasets/Iris.

at j. Otherwise, $x_{ij} = 0$ means no connection along this edge. Therefore, the cities form the set V of vertices and connections form the set E of the edges.

Let d_{ij} be the distance between city i and city j. Due to symmetry, we know that $d_{ij} = d_{ji}$, which means the graph is undirected. The objective is to

$$\text{minimize} \sum_{i,j \in E, i \neq j} d_{ij} x_{ij}. \tag{6.20}$$

One of the constraints is that

$$x_{ij} = 0 \quad \text{or} \quad 1, \quad (i, j) \in V. \tag{6.21}$$

Since each city i should be visited once and only once, it is required that

$$\sum_{i \in V} x_{ij} = 2, \quad \forall j \in V. \tag{6.22}$$

This means that only one edge enters a city and only one edge leaves the city, which can be equivalently written as

$$\sum_{j=1}^{n} x_{ij} = 1, \quad (i \in V, i \neq j), \tag{6.23}$$

and

$$\sum_{i=1}^{n} x_{ij} = 1, \quad (j \in V, j \neq i). \tag{6.24}$$

In order to avoid unconnected subtour, for any non-empty subset $S \subset V$ of n cities, it is required that

$$\sum_{i,j \in S} x_{ij} \leq |S| - 1, \quad (2 \leq |S| \leq n - 2), \tag{6.25}$$

where $|S|$ means the number of elements or the cardinality of the subset S. This is a binary integer linear programming problem, but it is NP-hard.

Due to the hardness of such problems, nature-inspired algorithms can be a good alternative. For example, Ouaarab et al. [75] used a discrete cuckoo search algorithm to solve various benchmark instances of the well-known travelling salesman problem, and they obtained the same or better results (about 2/3 of cases that are better), in comparison with the results obtained by other methods in the literature.

In addition, Osaba et al. [72] solved a class of symmetric and asymmetric travelling salesman problem by the improved bat algorithm, and they got very good results.

6.6 Vehicle Routing

Vehicle routing problems are a class of problems that are closely related to the travelling salesman problem. Vehicle routing can be complicated further by many factors such as traffic conditions, logistic requirements, distribution time windows (such as evening newspapers) and others. For example, Osaba et al. [73] used a discrete firefly algorithm to solve a type of vehicle routing problems for a newspaper distribution system with recycle policy in a provincial area in Spain, and they concluded that their discrete firefly algorithm outperformed the other two methods (evolutionary algorithm and evolutionary simulated annealing).

Furthermore, Osaba et al. [74] also solved a vehicle routing problem concerning medical goods distribution with pharmacological waste collection policy using a discrete and improved bat algorithm, and they achieved good results.

6.7 Scheduling

Scheduling itself is a class of challenging problems that can be NP-hard in many applications. Such scheduling problems can also be formulated as LP problems, though the detailed formulation can be problem-dependent.

One type of scheduling problems is the flow shop scheduling problem, which is most relevant to manufacturing industries such as steel, textile, electronics, chemical and pharmaceutical. Hybrid flow shop scheduling is more challenging to solve for allocating resources over time for a given set of tasks with multistage and parallel processors. Marichelvam et al. [66] used a discrete firefly algorithm to solve a multiobjective hybrid flow shop scheduling problem, and they showed that the discrete firefly algorithm performed better than genetic algorithm, ant colony optimization and simulated annealing.

In addition, Marichelvam et al. [67] also used an improved cuckoo search (ICS) to solve hybrid flow shop problems to minimize the overall makespan, and they concluded that ICS obtained better results than those by genetic algorithm, particle swarm optimization, ant colony optimization and other heuristics.

For other complicated and realistic scheduling problems, Pei et al. have recently carried out some detailed studies [80–83], concerning serial-batching scheduling with various constraints with extensive results.

6.8 Software Testing

Software should fit for the purpose of its functionalities, yet it should be reliable, robust and scalable, ideally without bugs. However, such reliability and robustness cannot be achieved without extensive testing and refinement. Thus, software testing is a crucial and complex part of the software development life cycle, and different tests should use different independent test paths. However, the generation of independent test paths is still a major challenge. It is estimated that about 50–60% of the development efforts may be dedicated to software testing.

Srivatsava et al. [94] used a discrete firefly algorithm (DFA) to generate optimal test sequences for software testing and they concluded that the DFA obtained faster results than ant colony optimization.

In addition, Srivastava et al. provide an approach for estimating the test efforts by using the cuckoo search algorithm [93]. For the various software projects under testing, they find that the cuckoo search based approach produced the most accurate estimation with a relative error under about 2%.

6.9 Deep Belief Networks

Deep learning is a very active area of research in artificial intelligence. For deep belief neural networks used for such deep machine learning, there are some hyper-parameters that need to be tuned. The settings of these parameters can affect the performance of such networks to a large extent, and fine-tuning is a time-consuming process. At the moment, most researchers used some empirical values based on experience, existing literatures and parametric studies. Ideally, some automatic fine-tuning by an optimization tool would be useful.

Papa et al. [78] used a quaternion representation in combination with a meta-heuristic approach to tune a restricted Boltzmann machine. Tests over the hand-written digits recognition showed promising results.

On the other hand, Rosa et al. [90] used various nature-inspired algorithms to optimize and handle the dropout probability for convolution neural networks for deep learning. Their tests concluded that most algorithms such as the bat algorithm (BA), particle swarm optimization and cuckoo search can obtain highly accurate results. In many cases, BA obtained the best and most accurate results for Semeion data set and USPS data set of hand-written digits.

6.10 Swarm Robots

Swarm intelligence is nowadays becoming increasingly effective and popular. It is not only a branch of theoretical studies and algorithms but also a reality of implementation in terms of software and hardware [77, 97].

For example, a swarm of robots can be used for rescue mission related to fire, chemical, biological hazardous environments or even for clearing of land mines to make the environment safer for humans. Robots inside a swarm communicate locally with a maximum range of influence. They interact according to local rules, without any central control. Such a system of swarming robots can potentially self-organize and form coalition to complete certain tasks. Thus, the balance of exploration and task completion can be a multiobjective problem. In recent studies by Palmieri et al. [77] and De Rango et al. [30] concerning a scenario of a large area with randomly distributed swarming robots and randomly distributed, unknown targets (see Figure 6.1), they showed that a hybrid approach of combining the firefly algorithm with the ant colony optimization can achieve very good results.

On the other hand, some researchers have developed special robots to mimic the true nature of natural systems and nature-inspired algorithms. For example, Suárez et al. [97] developed the bat robots with the hardware implementation that is specialized to mimic the exact nature of the bat algorithm. Their system of robots

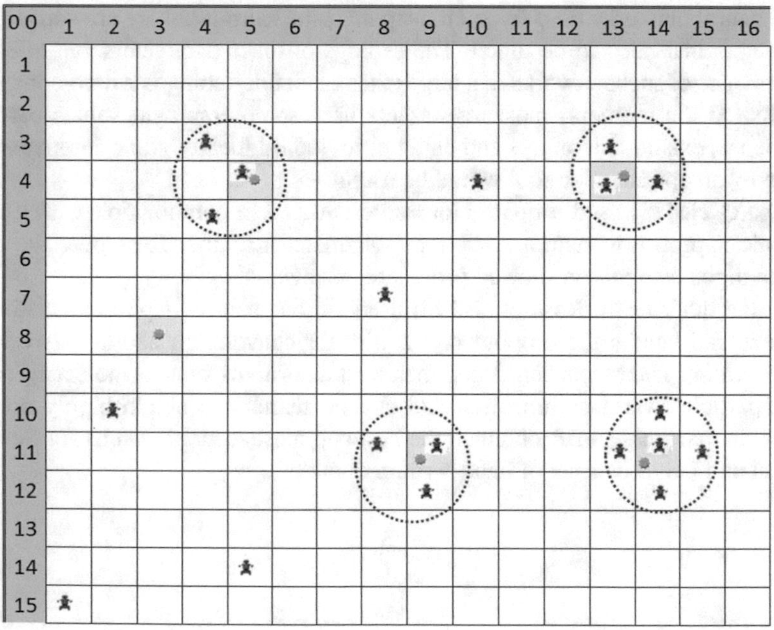

Fig. 6.1 A swarm of robots, interacting and collaborating by local rules [77]

can explore unknown regions, solving puzzles and potentially carry out rescue missions in real-world applications.

As we have seen in this chapter, there are a diverse range of applications concerning nature-inspired algorithms. The ten areas we mentioned are just a tiny fraction of the vast literature and applications, and the literature is rapidly expanding. Interested readers can refer to more specialized journal articles. We hope that this book can inspire more research in this rich and hot research area with more emerging real-world applications.

References

1. Abdelaziz AY, Ali ES, Abd Elazim SM (2016) Combined economic and emission dispatch solution using flower pollination algorithm. Int J Electr Power Energy Syst 80(2):264–274
2. Alam DF, Yousri DA, Eteiba MB (2015) Flower pollination algorithm based solar PV parameter estimation. Energy Convers Manag 101(2):410–422
3. Allan M (1977) Darwin and his flowers. Faber & Faber, London
4. Altringham JD (1998) Bats: biology and behaviour. Oxford University Press, Oxford
5. Alyasseri ZAA, Khader AT, Al-Betar MA, Awadallah MA, Yang XS (2017) Variants of the flower pollination algorithm: a review. In: Yang X-S (ed) Nature-inspired algorithms and applied optimization. Springer, Cham, pp 91–119
6. Arara S, Barak B (2009) Computational complexity: a modern approach. Cambridge University Press, Cambridge
7. Arora JS (1989) Introduction to optimum design. McGraw-Hill, New York
8. Ashby WA (1962) Principles of the self-organizing system. In: Von Foerster H, Zopf GW Jr (eds) Principles of self-organization: transactions of the University of Illinois Symposium. Pergamon Press, London, pp 255–278
9. Beer D (2016) The social power of algorithms. Inf Commun Soc 20(1):1–13
10. Bekdas G, Nigdeli SM, Yang XS (2015) Sizing optimization of truss structures using flower pollination algorithm. Appl Soft Comput 37(1):322–331
11. Bekdas G, Nigdeli SM, Yang XS (2018) A novel bat algorithm based optimum tuning of mass dampers for improving the seismic safety of structures. Eng Struct 159(1):89–98.
12. Bell WJ (1991) Searching behaviour: the behavioural ecology of finding resources. Chapman & Hall, London
13. Berlinski D (2001) The advent of the algorithm: the 300-year journey from an idea to the computer. Harvest Book, New York
14. Bertsekas DP, Nedic A, Ozdaglar A (2003) Convex analysis and optimization, 2nd edn. Athena Scientific, Belmont
15. Blum C, Roli A (2003) Metaheuristics in combinatorial optimization: overview and conceptual comparison. ACM Comput Survey 35(2):268–308
16. Bolton B (1995) A new general catalogue of the ants of the world. Harvard University Press, Cambridge
17. Bottou L (2010) Large-scale machine learning with stochastic gradient descent. In: Lechevallier Y, Saporta G (eds) Proceedings of COMPSTAT'2010, pp 177–186
18. Boyd S, Vandenberghe L (2004) Convex optimization. Cambridge University Press, Cambridge

X.-S. Yang, X.-S. He, *Mathematical Foundations of Nature-Inspired Algorithms*,
SpringerBriefs in Optimization, https://doi.org/10.1007/978-3-030-16936-7

19. Brin S, Page L (1998) The anatomy of a large-scale hypertextual web search engine. Comput Netw ISDN Syst 30(1–7):107–117
20. Cagnina LC, Esquivel SC, Coello CAC (2008) Solving engineering optimization problems with the simple constrained particle swarm optimizer. Informatica 32:319–326
21. Chabert JL (1999) A history of algorithms: from the pebble to the microchip. Springer, Heidelberg
22. Chakri A, Khelif R, Benouaret M, Yang XS (2017) New directional bat algorithm for continuous optimization problems. Expert Syst Appl 69(1):159–175
23. Chakri A, Yang XS, Khelif R, Benouaret M (2018) Reliability-based design optimization using the directional bat algorithm. Neural Comput Appl 30(8):2381–2402
24. Chen S, Peng G-H, He XS, Yang XS (2018) Global convergence analysis of the bat algorithm using a Markovian framework and dynamic system theory. Expert Syst Appl 114(1):173–182
25. Clerc M, Kennedy J (2002) The particle swarm - explosion, stability, and convergence in a multidimensional complex space. IEEE Trans Evol Comput 6(1):58–73
26. Coello CAC (2000) Use of a self-adaptive penalty approach for engineering optimization problems. Comput Ind 41(2):113–127
27. Copeland BJ (2004) The essential Turing. Oxford University Press, Oxford
28. Dantzig GB, Thapa MN (1997) Linear programming. 1: Introduction. Springer, Heidelberg
29. Davies NB (2011) Cuckoo adaptations: trickery and tuning. J Zool 284(1):1–14
30. De Rango F, Palmieri N, Yang XS, Marano S (2018) Swarm robotics in wireless distributed protocol design for coordinating robots involved in cooperative tasks. Soft Comput 22(13):4251–4266
31. Dorigo M (1992) Optimization, learning and natural algorithms. Ph.D. Thesis, Politecnico di Milano, Italy
32. Eiben AE, Smit SK (2011) Parameter tuning for configuring and analyzing evolutionary algorithms. Swarm Evol Comput 1(1):19–31
33. Engelbrecht AP (2005) Fundamentals of computational swarm intelligence. Wiley, Hoboken
34. Fisher L (2009) The perfect swarm: the science of complexity in everyday life. Basic Books, New York
35. Fishman GS (1995) Monte Carlo: concepts, algorithms and applications. Springer, New York
36. Fister I, Yang XS, Brest J, Fister I Jr (2013) Modified firefly algorithm using quaternion representation. Expert Syst Appl 40(18):7220–7230
37. Fogel LJ, Owens AJ, Walsh MJ (1966) Artificial intelligence through simulated evolution. Wiley, New York
38. Fong S, Zhuang Y, Tang R, Yang XS, Deb S (2013) Selecting optimal feature set in high-dimensional data by swarm search. J Appl Math 2013:18 pp., Article ID: 590614. http://dx.doi.org/10.1155/2013/590614
39. Freedman DA (2009) Statistical models: theory and practice. Cambridge University Press, Cambridge
40. Gandomi AH, Yang XS (2014) Chaotic bat algorithm. J Comput Sci 5(2):224–232
41. Gandomi AH, Yang XS, Alavi AH (2013) Cuckoo search algorithm: a metaheuristic approach to solve structural optimization problems. Eng Comput 29(1):17–35
42. Geem ZW, Kim JH, Loganathan GV (2001) A new heuristic optimization algorithm: harmony search. Simulation 76(2):60–68
43. Ghate A, Smith R (2008) Adaptive search with stochastic acceptance probabilities for global optimization. Oper Res Lett 36(3):285–290
44. Glover F (1986) Future paths for integer programming and links to artificial intelligence. Comput Oper Res 13(5):533–549
45. Glover BJ (2007) Understanding flowers and flowering: an integrated approach. Oxford University Press, Oxford
46. Goldberg DE (1989) Genetic algorithms in search, optimisation and machine learning. Addison Wesley, Reading
47. Grindstead CM, Snell JL (1997) Introduction to probability, 2nd edn. American Mathematical Society, Providence

48. He XS, Yang XS, Karamanoglu M, Zhao YX (2017) Global convergence analysis of the flower pollination algorithm: a discrete-time Markov chain approach. Proc Comput Sci 108(1):1354–1363
49. Holland J (1975) Adaptation in natural and artificial systems. University of Michigan Press, Ann Arbor
50. Hölldobler B, Wilson EO (2009) The superorganism: the beauty, elegance and strangeness of insect societies. Norton & Co, New York
51. Huber PJ (1981). Robust statistics. Wiley, New York
52. Jamil M, Yang XS (2013) A literature survey of benchmark functions for global optimisation problems. Int J Math Model Numer Optim 4(2):150–194
53. Joyce T, Herrmann JM (2018) A review of no free lunch theorems, and their implications for metaheuristic optimisation. In: Yang XS (ed) Nature-inspired algorithms and applied optimization. Springer, Berlin, pp 27–52
54. Judea P (1984) Heuristics. Addison-Wesley, New York
55. Karaboga D (2005) An idea based on honeybee swarm for numerical optimization. Technical Report, Erciyes University, Turkey
56. Keller EF (2009) Organisms, machines, and thunderstorms: a history of self-organization, Part two. Complexity, emergence, and stable attractors. Hist Stud Nat Sci 39(1):1–31
57. Kennedy J, Eberhart RC (1995) Particle swarm optimization. In: Proceedings of IEEE international conference on neural networks, Piscataway, pp 1942–1948
58. Kennedy J, Eberhart RC, Shi Y (2001) Swarm intelligence. Academic Press, London
59. Kirkpatrick S, Gellat CD, Vecchi MP (1983) Optimization by simulated annealing. Science 220(4598):671–680
60. Koza JR (1992) Genetic programming: on the programming of computers by means of natural selection. MIT Press, Cambridge
61. Lazer D (2015) The rise of the social algorithm. Science 348(6239):1090–1091
62. LeCun Y, Bengio Y, Hinton GE (2015). Deep learning. Nature 521:636–444
63. Lewis SM, Cratsley CK (2008) Flash signal evolution, mate choice and predation in fireflies. Annu Rev Entomol 53(2):293–321
64. Lindauer M (1971) Communication among social bees. Harvard University Press, Cambridge
65. Mantegna RN (1994) Fast, accurate algorithm for numerical simulation of Lévy stable stochastic processes. Phys Rev E 49:4677–4683
66. Marichelvam MK, Prabaharan T, Yang XS (2014) A discrete firefly algorithm for multi-objective hybrid flowshop scheduling problems. IEEE Trans Evol Comput. 18(2):301–205
67. Marichelvam MK, Prabaharan T, Yang XS (2014) Improved cuckoo search algorithm for hybrid flowshop scheduling problems to minimize makespan. Appl Soft Comput 19(1):93–101
68. Moravej Z, Akhlaghi A (2013) A novel approach based on cuckoo search for DG allocation in distribution network. Electr Power Energy Syst 44(1):672–679
69. Musselman RD (2007) Robustness: a better measure of algorithm performance. Master Thesis, Naval Postgraduate School, Monterey
70. Nakamura RYM, Fonseca LMG, dos Santos JA, Torres RS, Yang XS, Papa JP (2014) Nature-inspired framework for hyperspectral bank selection. IEEE Trans Geosci Remote Sens 52(4):2126–2137
71. Nakrani S, Tovey C (2004) On honeybees and dynamic server allocation in Internet hosting centers. Adapt Behav 12(3):223–240
72. Osaba E, Yang XS, Diaz F, Lopez-Garcia P, Carballedo R (2016) An improved discrete bat algorithm for symmetric and asymmetric travelling salesman problems. Eng Appl Artif Intell 48(1):59–71
73. Osaba E, Yang XS, Diaz F, Onieva E, Masegosa AD, Perallos A (2017) A discrete firefly algorithm to solve a rich vehicle routing problem modelling a newspaper distribution system with recycling policy. Soft Comput 21(18):5295–5308

74. Osaba E, Yang XS, Fister I Jr, Del Ser J, Lopez-Garcia P, Vazquez-Pardavila AJ (2019) A discrete and improved bat algorithm for solving a medical goods distribution problem with pharmacological waste collection. Swarm Evol Comput 44(1):273–286
75. Ouaarab A, Ahiod B, Yang XS (2014) Discrete cuckoo search algorithm for the travelling salesman problem. Neural Comput Appl 24(7–8):1659–1669
76. Page L, Brin S, Motwani R, Winograd T (1998) The pagerank citation ranking: bringing order to the web. Technical Report, Stanford University
77. Palmieri N, Yang XS, De Rango F, Santamaria AF (2018) Self-adaptive decision-making mechanisms to balance the execution of multiple tasks for a multi-robots team. Neurocomputing 306(1):17–36
78. Papa JP, Rosa GH, Pereira DR, Yang XS (2017) Quaternion-based deep belief networks fine-tuning. Appl Soft Comput 60:328–335
79. Pavlyukevich I (2007) Lévy flights, non-local search and simulated annealing. J Comput Phys 226(2):1830–1844
80. Pei J, Pardalos PM, Liu XB, Fan WJ, Yang SL (2015) Serial batching scheduling of deteriorating jobs in a two-stage supply chain to minimize the makespan. Eur J Oper Res 244(1):13–25
81. Pei J, Darzic Z, Drazic M, Mladenovic N, Pardalos P (2018) Continuous variable neighborhood search (C-VNS) for solving systems of nonlinear equations. INFORMS J Comput. https://doi.org/10.1287/ijoc.2018.0876
82. Pei J, Cheng BY, Liu XB, PM Pardalos, Kong M (2019) Single-machine and parallel-machine serial-batching scheduling problems with position-based learning effect and linear setup time. Ann Oper Res 272(1–2):217–241
83. Pen J, Liu XB, Fang WJ, Pardolas PM, Lu SJ (2019) A hybrid BA-VNS algorithm for coordinated serial-batching scheduling with deteriorating jobs, financial budget, and resource constraint in multiple manufacturers. Omega 82(1):55–69
84. Pereira LAM, Rodrigues D, Almeida TNS, Ramos CCO, Souza AN, Yang XS, Papa JP (2014) A binary cuckoo search and its application for feature selection. In: Yang XS (ed) Cuckoo search and firefly algorithm, pp 141–154
85. Pham DT, Ghanbarzadeh A, Koc E, Otri S, Rahim S, Zaidi M (2005) The bees algorithm, Technical Note, Manufacturing Engineering Centre, Cardiff University
86. Press WH, Teukolsky SA, Vetterling WT, Flannery BP (2007) Numerical recipes: the art of scientific computing, 3rd edn. Cambridge University Press, Cambridge
87. Rashedi E, Nezamabadi-pour H, Sayazdi S (2009) GSA: a gravitational search algorithm. Inf Sci 179(13):2232–2248
88. Reynolds AM, Rhodes CJ (2009) The Lévy flight paradigm: random search patterns and mechanisms. Ecology 90(4):877–887
89. Rodrigues D, Silva GFA, Papa JP, Marana AN, Yang XS (2016) EEG-based person identification through binary flower pollination algorithm. Expert Syst Appl 62(1):81–90
90. Rosa GH, Papa JP, Yang XS (2017) Handling dropout probability estimation in convolutional neural networks using meta-heuristics. Soft Comput. https://doi.org/10.1007/s00500-017-2678-4
91. Senthilnath J, Yang XS, Benediktsson JA (2014) Automatic registration of multi-temporal remote sensing images based on nature-inspired techniques. Int J Image Data Fusion 5(4):263–284
92. Senthilnath J, Kulkarni S, Benediktsson JA, Yang XS (2016) A novel approach for multi-spectral satellite image classification based on the bat algorithm. IEEE Geosci Remote Sens Lett 13(4):599–603
93. Srivastava PR, Varshney A, Nama P (2012) Software test effort estimation: a model based on cuckoo search. Int J Bio-Inspired Comput 4(5):278–285
94. Srivatsava PR, Millikarjun B, Yang XS (2013) Optimal test sequence generation using firefly algorithm. Swarm Evol Comput 8(1):44–53
95. Storn R, Price K (1997) Differential evolution: a simple and efficient heuristic for global optimization over continuous spaces. J Global Optim 11(4):341–359

96. Struik DJ (1987) A concise history of mathematics, 4th edn. Dover, New York
97. Suárez P, Iglesias A, Gálvez A (2018) Make robots be bats: specializing robotic swarm to the bat algorithm. Swarm Evol Comput, Published online 14 Feb 2018. https://doi.org/10.1016/j.swevo.2018.01.005
98. Süli E, Mayer D (2003) An introduction to numerical analysis. Cambridge University Press, Cambridge
99. Surowiecki J (2004) The wisdom of crowds. Doubleday, Anchor
100. Suzuki JA (1995) A Markov chain analysis on simple genetic algorithms. IEEE Trans Syst Man Cybern 25(4):655–659 (1995)
101. Taguchi G (1987) System of experimental design: engineering methods to optimize quality and minimize costs, vols. 1 and 2. ASI Press, New York
102. Taguchi G, Jugulum R, Taguchi S (2005) Computer-based robust engineering. American Society of Quality (ASQ) Quality Press, Milwaukee
103. Turing AM (1948) Intelligent machinery. National Physical Laboratory, Technical Report, Teddington
104. Vapnik V (1995) The nature of statistical learning theory. Springer, New York
105. Villalobos-Arias M, Colleo CAC, Hernández-Lerma O (2005) Asymptotic convergence of metaheuristics for multiobjective optimization problems. Soft Comput 10(11):1001–1005
106. Waser NM (1986) Flower constancy: definition, cause and measurement. Am Nat 127(5):596–603
107. Wolpert DH, Macready WG (1997) No free lunch theorem for optimization. IEEE Trans Evol Comput 1(1):67–82
108. Wolpert DH, Macready WG (2005) Coevolutionary free lunches. IEEE Trans Evol Comput 9(6):721–735
109. Yang XS (2005) Engineering optimization via nature-inspired virtual bee algorithms. In: Artificial intelligence and knowledge engineering application: a bioinspired approach, proceedings of IWINAC, pp 317–323
110. Yang XS (2008) Nature-inspired metaheuristic algorithms. Luniver Press, Bristol
111. Yang XS (2010) Firefly algorithm, stochastic test functions and design optimisation. Int J Bio-Inspired Comput 2(2):78–84
112. Yang XS (2010) A new metaheuristic bat-inspired algorithm. In: Nature-inspired cooperative strategies for optimization (NICSO 2010). Studies in computational intelligence, vol 284. Springer, Berlin, pp 65–74
113. Yang XS (2010) Engineering optimization: an introduction with metaheuristic applications. Wiley, Hoboken
114. Yang XS (2011) Metaheuristic optimization. Scholarpedia 6(8):11472
115. Yang XS (2011) Bat algorithm for multi-objective optimisation. Int J Bio-Inspired Comput 3(5):267–274
116. Yang XS (2012) Flower pollination algorithm for global optimization. In: Unconventional computation and natural computation. Lecture notes in computer science, vol 7445, pp 240–249
117. Yang XS (2014) Cuckoo search and firefly algorithm: theory and applications. Studies in computational intelligence, vol 516. Springer, Heidelberg
118. Yang XS (2014) Nature-inspired optimization algorithms. Elsevier Insight, London
119. Yang XS (2017) Engineering mathematics with examples and applications. Academic, London
120. Yang XS (2018) Nature-inspired algorithms and applied optimization. Springer, Cham
121. Yang XS (2018) Optimization techniques and applications with examples. Wiley, Hoboken
122. Yang XS, Deb S (2009) Cuckoo search via Lévy flights. In: Proceedings of world congress on nature & biologically inspired computing (NaBic 2009), Coimbatore, India. IEEE, Piscataway, pp 210–214
123. Yang XS, Deb S (2010) Engineering optimization by cuckoo search. Int J Math Model Num Optim 1(4):330–343

124. Yang XS, Deb S (2013) Multiobjective cuckoo search for design optimization. Comput Oper Res 40(6):1616–1624
125. Yang XS, Deb S (2014) Cuckoo search: recent advances and applications. Neural Comput Appl 24(1):169–174
126. Yang XS, Gandomi AH (2012) Bat algorithm: a novel approach for global engineering optimization. Eng Comput 29(5):464–483
127. Yang XS, He XS (2013) Bat algorithm: literature review and applications. Int J Bio-Inspired Comput 5(3):141–149
128. Yang XS, Papa JP (2016) Bio-inspired computation and applications in image processing. Academic Press, London
129. Yang XS, Huyck CR, Karamanoglu M, Khan N (2013) True global optimality of the pressure vessel design problem: a benchmark for bio-inspired optimisation algorithms. Int J Bio-Inspired Comput 5(6):329–335
130. Yang XS, Deb S, Loomes M, Karamanoglu M (2013) A framework for self-tuning optimization algorithm. Neural Comput Appl 23(7–8):2051–2057
131. Yang XS, Cui ZH, Xiao RB, Gandom AH, Karamanoglu M (2013) Swarm intelligence and bio-inspired computation: theory and applications. Elsevier, London
132. Yang XS, Karamanoglu M, He XS (2014) Flower pollination algorithm: a novel approach for multiobjective optimization. Eng Optim 46(9):1222–1237
133. Yang XS, Chien SF, Ting TO (2015) Bio-inspired computation in telecommunications. Morgan Kaufmann, Waltham
134. Yang XS, Deb S, Mishra SK (2018) Multi-species cuckoo search for global optimization. Cogn Comput 10(6):1085–1095
135. Yildiz AR (2013) Cuckoo search algorithm for the selection of optimal machine parameters in milling operations. Int J Adv Manuf Technol 64(1):55–61
136. Zaharie D (2009) Influence of crossover on the behaviour of the differential evolution algorithm. Appl Soft Comput 9(3):1126–1138

Index

A
Accelerated PSO, 32
Algorithm, 2, 59
Annealing, 25
Annealing schedule, 25
Ant colony optimization, 22, 28
Arithmetic complexity, 66

B
Bat algorithm, 23, 34, 68
Bayesian framework, 76
Bees-inspired algorithm, 22, 33
Biased Monte Carlo, 70
Bio-inspired algorithms, 24
Boltzmann distribution, 26

C
Classification, 92
Clustering, 92
Co-evolution, 38
Co-evolutionary algorithm, 85
Complexity, 64
Conjugate gradient method, 16
Constrained optimization, 10
Convergence, 12, 41
Convergence analysis, 41, 42
Convex, 5
Cooling process, 26
Cooling rate, 25
Crossover, 23, 29
  binomial, 29
  exponential, 29
Cuckoo search, 23, 38
Cumulative probability function, 49

D
Deep belief network, 95
Design optimization, 87
Deterministic, 62
DE variant, 30
Differential evolution, 22, 29
Dynamical system, 67

E
Evolutionary strategy, 21
Exploitation, 62
Exploration, 62
Exponential distribution, 50

F
Feature selection, 92
Filter theory, 76
Firefly algorithm, 22, 35, 83
Fixed point theory, 66
Flower pollination algorithm, 23, 39

G
Gaussian distribution, 48
Genetic algorithm, 21, 23, 24
Global optimality, 27
Global search, 61
Gradient-based method, 11

H
Hessian matrix, 12
Heuristic, 61

X.-S. Yang, X.-S. He, *Mathematical Foundations of Nature-Inspired Algorithms*, SpringerBriefs in Optimization, https://doi.org/10.1007/978-3-030-16936-7

Hill-climbing method, 15
Hybrid algorithm, 85
Hyper-optimization, 81

I
Image processing, 91
Inertia function, 32
Initial temperature, 26
Inverse problem, 90
Iteration, 12

L
Lévy flight, 52
Linear program, 10
Line search, 15
Local search, 61

M
Mantegna algorithm, 54
Markov chain, 56, 69
Mathematical analysis, 59, 75
Maximum, 5
Mean, 47
Metaheuristic, 22, 57, 61
Minimum, 5
Minimum energy, 25
Modified Newton's method, 12
Monte Carlo, 56, 69
Multimodal, 6
Multimodal function, 6
Multiobjective, 23
Mutation, 23, 29

N
Nature-inspired algorithm, 10, 21, 57, 59, 87
Newton's method, 3, 11
No-free-lunch theorem, 22, 72
Normal distribution, 48

O
Objective, 11
Optimization, 1, 10, 29, 87
  multivariate, 7
  univariate, 4

P
Parameter control, 78, 80
Parameter identification, 90
Parameter tuning, 78, 79, 84
Pareto distribution, 52
Particle swarm optimization (PSO), 22, 31
Performance measure, 54
Poisson distribution, 48
Power-law distribution, 51
Pressure vessel design, 88
Probability
  crossover, 24
  distribution, 47
  mutation, 24
Probability density function, 48
PSO, *see* Particle swarm optimization

Q
Quasi-Newton method, 13

R
Random search, 26
Random variable, 45
Random walk, 52, 71
Rate of convergence, 41
Robustness analysis, 45
Role of components, 63

S
SA, *see* Simulated annealing
Scheduling, 94
Selection, 23, 29
Self-adaptivity, 86
Self-evolving algorithm, 86
Self-organization, 60
Self-organized system, 68
Self-tuning algorithm, 83
Self-tuning framework, 82
Simulated annealing (SA), 21, 25
Software testing, 95
Space complexity, 64
Stability, 27, 44
Stationary point, 6, 8
Steepest descent method, 13
Stochastic, 62
Stochastic gradient descent, 18

Stochastic learning, 77
Subgradient method, 19, 20
Swarm intelligence, 75
Swarm robots, 96

T
Time complexity, 65
Travelling salesman problem, 23, 92
Turing machine, 66

U
Unconstrained optimization, 4
Uniform distribution, 49

V
Variance, 47
Variant, 30
Vehicle routing, 94